Towards Disaster Risk Reduction

This book constitutes a landmark attempt to address, comprehensively and in depth, a policy-focused approach to the many timely and important issues associated with building a culture of disaster prevention and disaster risk reduction. This book not only provides key insights into the field of natural hazard and disaster studies but also assesses the causes, perspectives, and consequences of natural disasters, as well as providing a global survey of post-recovery policies. The contributions found herein discuss disaster risk reduction strategies and policies for managing the unexpected and cascading impacts of natural disasters. A particular focus is placed on transboundary catastrophes that cross policy domains, geographic, political, and sectoral boundaries. Since the disaster management and natural resources policy research field draws on a diverse range of paradigms and influences, the book includes case histories, empirical studies, conceptual-theoretical investigations, policy perspectives, institutional analysis, and risk analyses. The role of human culture, disaster psychology and environmental monitoring are examined in depth. Deficiencies and inequalities in local, national, and global disaster response are also discussed. Original strategies for reducing disaster risk and the prospects for a major change in the direction of global policy on disasters are put forward.

This book was published as a special issue of the *Journal of Natural Resources Policy Research*.

Jason K. Levy is Associate Professor of Homeland Security and Emergency Preparedness at Virginia Commonwealth University, USA. He is also Director of the National Homeland Security Project.

Chennat Gopalakrishnan is Professor at the Department of Natural Resources and Environmental Management, University of Hawaii at Manoa, US.

Towards Disaster Risk Reduction
A Policy-Focused Approach to Natural Hazards and Disasters

Edited by
Jason K. Levy and Chennat Gopalakrishnan

Routledge
Taylor & Francis Group
LONDON AND NEW YORK

First published 2013
by Routledge
2 Park Square, Milton Park, Abingdon, Oxon, OX14 4RN

Simultaneously published in the USA and Canada
by Routledge
711 Third Avenue, New York, NY 10017

Routledge is an imprint of the Taylor & Francis Group, an informa business

This book is a reproduction of the *Journal of Natural Resources Policy Research*, vol. 2/issue 4. The Publish
requests to those authors who may be citing this book to state, also, the bibliographical details of the speci
issue on which the book was based.

British Library Cataloguing in Publication Data
A catalogue record for this book is available from the British Library

ISBN13: 978-0-415-82678-5

Typeset in Times New Roman
by Taylor & Francis Books

Publisher's Note
The publisher would like to make readers aware that the chapters in this book may be referred to as article
as they are identical to the articles published in the special issue. The publisher accepts responsibility for an
inconsistencies that may have arisen in the course of preparing this volume for print.

Contents

Citation Information vii

1. Introduction: A Policy-Focused Approach to Natural Hazards and Disasters - Towards Disaster Risk Reduction (DRR)
Jason K. Levy & Chennat Gopalakrishnan 1

2. The L'Aquila Earthquake of 6 April 2009 and Italian Government Policy on Disaster Response
David E. Alexander 8

3. Using Natural Capital to Reduce Disaster Risk
Carolyn Kousky 26

4. The Dynamics of Informal Responses to Covariate Shocks
R. A. Balgah & G. Buchenrieder 40

5. Negotiation Support for Environmental Disaster Management: Drama Theory II and the Survival of Louisiana's Indigenous Gulf Coast Tribes
Jason K. Levy 54

6. A Gap Hazard Analysis: Initiating Policy Development with Mountainous Communities
G. Strickert, S. Smarasinghe, C. Doscher & T. Davies 72

Index 91

Citation Information

The chapters in this book were originally published in the *Journal of Natural Resources Policy Research*, volume 2, issue 4 (October 2010). When citing this material, please use the original page numbering for each article, as follows:

Chapter 1
Introduction: A Policy-Focused Approach to Natural Hazards and Disasters - Towards Disaster Risk Reduction (DRR)
Jason K. Levy & Chennat Gopalakrishnan
Journal of Natural Resources Policy Research, volume 2, issue 4 (October 2010) pp. 317-324

Chapter 2
The L'Aquila Earthquake of 6 April 2009 and Italian Government Policy on Disaster Response
David E. Alexander
Journal of Natural Resources Policy Research, volume 2, issue 4 (October 2010) pp. 325-342

Chapter 3
Using Natural Capital to Reduce Disaster Risk
Carolyn Kousky
Journal of Natural Resources Policy Research, volume 2, issue 4 (October 2010) pp. 343-356

Chapter 4
The Dynamics of Informal Responses to Covariate Shocks
R. A. Balgah & G. Buchenrieder
Journal of Natural Resources Policy Research, volume 2, issue 4 (October 2010) pp. 357-370

Chapter 5
Negotiation Support for Environmental Disaster Management: Drama Theory II and the Survival of Louisiana's Indigenous Gulf Coast Tribes
Jason K. Levy
Journal of Natural Resources Policy Research, volume 2, issue 4 (October 2010) pp. 371-388

Chapter 6
A Gap Hazard Analysis: Initiating Policy Development with Mountainous Communities
G. Strickert, S. Smarasinghe, C. Doscher & T. Davies
Journal of Natural Resources Policy Research, volume 2, issue 4 (October 2010) pp. 389-407

Introduction: A Policy-Focused Approach to Natural Hazards and Disasters - Towards Disaster Risk Reduction (DRR)

JASON K. LEVY* & CHENNAT GOPALAKRISHNAN**

*Homeland Security and Emergency Preparedness, L. Douglas Wilder School of Government, Virginia Commonwealth University, Richmond, VA, USA; **Department of Natural Resources and Environmental Management, University of Hawaii at Manoa, Honolulu, HI, USA

While the fields of disaster, crisis, and emergency studies have made great strides over the past century (Alexander, 1993; Mileti, 1999; Prince, 1920; Quarantelli, 1998; White, 1942), the international relief community is currently being pushed to the brink by the unparalleled intensity, frequency, and scale of natural disasters and catastrophes. In 2009 alone, 335 natural disasters occurred worldwide, killing 10 655 persons, disrupting the lives of 119 million others and causing over US$41.3 billion economic damages, with Asia accounting for 89.1% of global reported natural disaster victims and more than a third of total worldwide economic damages (Vos *et al.*, 2010). In 2005, Hurricane Katrina alone displaced more than a quarter of a million Gulf Coast citizens, damaged a land area equal to the United Kingdom, destroyed 300 000 homes, and cost $US200 billion (Duncan & Brebbia, 2009). Striking in the most populated area of the country, the 2010 Haiti earthquake affected approximately 3 million people with a reported death toll of 230 000 and 300 000 injured (BBC News, 2010). In 2008 alone there were 235 000 deaths due to natural disasters: cyclone Nagaris in Myanmar killed 138 366 people, while the Sichuan earthquake in China killed 87 476 people (Vos *et al.*, 2010).

Clearly, the death toll of recent 'mega-disasters' has been staggering. The United Nations has rated the August 2010 floods in Pakistan as the greatest humanitarian crisis in recent history, with more people affected than by the 2004 Boxing Day South-East Asian tsunami and the recent earthquakes in Kashmir and Haiti combined. UN Secretary-General Ban Ki-Moon has requested additional urgent aid to Pakistan because the floods have already affected an area roughly the size of Italy, disrupted the lives of a tenth of its 170 million population, left two million homeless, and killed approximately 1600, amid fears that ease outbreaks could spread since survivors are sleeping in makeshift tarpaulin tents. Th events highlight the fact that natural, health-related and technologic disasters are on rise worldwide, including statistically-probable but unanticipated catastrophes as wel moderate-scale repetitive events. Global climate variability and change, infectious dise epidemics and catastrophic deforestation pose additional long-term challenges to ecol cal integrity and global population health. For example, rising sea levels currently po threat to more than half a billion people that live within 5 m above sea level around world. Sea-level rise is an ongoing and accelerating process and the communities of s islands and deltaic coasts are unable to retreat inland from the coast. The number of gl 'environmental refugees' facing involuntary relocation is expected to reach 50 million 2010, with small, low-lying island populations at the greatest risk (Potter, 2008). For ex ple, in the Pacific nation of Tuvalu, a ring of nine Polynesian islands, several thous people have already left for other nations due to the loss of land.

The five papers in this special issue constitute a major attempt to address, compreh sively and in-depth, a policy-focused approach to the many timely and important iss associated with building a culture of disaster prevention and disaster risk reduction. T special issue emphasizes that devastation and loss allows for new opportunities to man disaster risk in ways characterized by adaptation, transformation and resilience: traditio plans and policies for mitigating disaster losses are inadequate and so decision-mak must understand how natural hazards impact their communities and the policy framewo in which they are managed. This special issue not only provides key insights into the fi of natural hazard and disaster studies but also assesses the causes, perspectives, and con quences of natural disasters, as well as providing a global survey of post-recovery polic The five contributions found herein discuss disaster risk reduction strategies and policies managing the unexpected and cascading impacts of natural disasters. A particular focu placed on transboundary catastrophes that cross policy domains, geographic, political, sectoral boundaries. Since the disaster management and natural resources policy resea field draws on a diverse range of paradigms and influences, the five papers featured incl case histories, empirical studies, conceptual-theoretical investigations, policy perspecti institutional analysis, and risk analyses.

This special issue investigates current gaps in our knowledge about managing nat hazards with the aim of opening new research avenues for integrated disaster risk mana ment, from a policy perspective. Accordingly, this collection includes cross-disciplin and transnational papers that examine the causes and consequences of natural disast with a special emphasis on crafting comprehensive disaster policy solutions to effectiv address them. Key topics discussed include: adaptive management of natural disast risk and vulnerability assessments, and the implementation of disaster mitigation p cies. Trends in the frequency and impacts of disasters in both developed and develop nations are examined and explanations for these calamities are advanced. The role human culture, disaster psychology, and environmental monitoring are examined in de Deficiencies and inequalities in local, national, and global disaster response are also c cussed. Original strategies for reducing disaster risk are put forward and the prospects a major change in the direction of world policy on disasters are assessed.

In the first paper, Alexander provides a comprehensive analysis, with general less extracted, of short- to medium-term disaster policies (i.e. regarding emergency mana ment, temporary resettlement, and initial stages of permanent recovery). Specifically, author describes the impact of the earthquake that struck the central Italian city of L'Aquila on 6 April 2009, killing 308 people and leaving 67 500 homeless. The pre-impact, emergency, and early recovery phases are discussed in terms of the nature and effectiveness of government policy. Disaster risk reduction in Italy is evaluated in relation to the structure of civil protection and changes wrought by both the L'Aquila disaster and public scandals connected with the misappropriation of funds. Six important lessons (see below) are derived from this analysis and related to disaster risk reduction (DRR) needs both in Italy and elsewhere in the world. Policy analysis is provided with respect to earthquake prediction in Italy. The value of short-term earthquake forecasts is discussed.

Alexander identifies six important policy lessons of the L'Aquila earthquake disaster. First, whether short-term earthquake predictions are official or amateur, the precautionary principle should be employed. Second, monitoring and warning systems must be backed by robust and local emergency response mechanisms. Third, it is necessary to augment, not supplant, local disaster response with external assistance. This requires an integrated system of emergency response. Fourth, safeguards against corruption and abuse need to be instituted; civil protection must necessarily have both recourse to emergency powers as well as democratic underpinnings. Fifth, in most disaster response and emergency management systems, there is a tension between centrism and devolution. Finally, the price of progress is eternal vigilance: disaster reduction gains made in one year can be reversed in another when budgets are cut and programmes are reduced.

In the next paper, Kousky discusses the use of natural capital to reduce disaster risk. The author emphasizes that natural capital can be used to reduce disaster losses (i.e. wetlands store floodwaters and mangroves buffer storm surge) and is hence an important disaster management tool. Over the past half century key contributions have been made to the fields of natural capital, poverty, and disaster risk (Wisner *et al.*, 1976), the social sciences and natural hazards (Wright & Rossi, 1981), collective behaviour and disaster research (Wenger, 1985), and the 'human ecological perspective' to hazards. To complement our growing scientific understanding of disaster probabilities and consequences, Kousky calls attention to the policy process of using natural capital over more traditional risk management approaches. Lower costs and co-benefits can attract policy-makers to use the natural capital approach, and a variety of policy mechanisms exist for achieving disaster risk reduction.

Kousky's paper reviews the enlightening experience of several vanguard US communities that are employing natural systems to mitigate disaster risks. The paper concludes with a discussion of the motivations for using the natural capital approach (as opposed to other risk reduction measures), the policy approaches for doing so, and the political and institutional challenges involved. Four potential challenges to adopting the use of natural capital for disaster risk reduction are identified: lack of information, uninterested decision-makers, political opposition, and institutional biases and inertia. Communities have effectively found ways around these challenges, however, by taking advantage of triggering events, working with dedicated public entrepreneurs, and turning to potentially new methods, such as multi-stakeholder negotiation.

In the third paper, Balagh and Buchenrieder discuss the resilience of formal and informal instruments to natural and man-made shocks which result in income or consumption volatility. For example, in the event of a drought event, affected households employ various formal and informal instruments with the aim of achieving responsive outcomes such as consumption smoothing (Holzmann *et al.* 2003) or asset smoothing (Zimmerman &

Carter 2003; Carter *et al.* 2007). Balagh and Buchenrieder note that informal arran ments describe the bundle of measures taken at the household or community level to pro against risks (or to combat shocks) in the absence or presence of public or market-ba arrangements. Examples include the buying and selling of real assets (such as cattle gold), informal borrowing, crop diversification, storage, and arrangements for commu support. Next, the authors describe formal instruments: market and public actions aga precedented and unprecedented risks or shocks. The term 'risk' is used when referr to uncertain (i.e. stochastic) events and outcomes with known or unknown probabili distributions (IFPRI, 2002; Sricharoen, 2006).

Using empirical evidence, the authors argue that clear evidence does not support superiority of formal instruments over informal responses to covariate shocks. To the c trary, the authors propose that the dynamic response of informal instruments to aggreg shocks is often misunderstood and underestimated. The paper concludes with suggesti on pertinent research issues and future work in order to better explain the role of in mal structures in managing shocks. The expected outcomes of this research include m robust disaster management plans capable of coping with a rapidly changing environm Balagh and Buchenrieder conclude by emphasizing the importance of promoting dis ter management education and incorporating the principles of disaster risk reduction social disaster resilience into the broader issues of ecosystems integrity and sustaina livelihoods.

In the next paper, Levy describes the theory and practice of drama theory for manag disasters and improving natural resources policy research. A formal group decision negotiation framework known as Drama Theory II (DT II) is put forth in order to impr environmental disaster modelling and analysis. The author investigates the natural, hea related, and man-made challenges facing Louisiana's indigenous Gulf Coast tribes that h lived on the bayous of the Mississippi delta for centuries. A mathematical treatment of drama-theoretic concepts (i.e. positions, intentions, doubts, and dilemmas) is provided, dynamics of the drama-theoretic process are discussed, and theoretical results are appl to the oil spill disaster in the US Gulf Coast. It is shown that DT II successfully mod emotional responses during a disaster in order to improve strategic decision-making reduce disaster risk.

Levy notes that there are several key distinctions between drama theory and game t ory with respect to disaster management: DT II makes the analysis more relevant applicable to real-world disasters by eliminating the need for the cardinal (von-Neum Morgenstern) payoffs and mixed (i.e. probabilistic) strategies used in game theory (dra theory uses ordinal preferences and pure strategies); drama theory assumes any numbe players (each of whom has a large number of options or alternatives to choose from); game-theoretic analysis identifies equilibria within a fixed model which requires play to use artificially fixed preferences and options (and to make decisions based on mo in a model calibrated at the beginning of the analysis). Levy notes the limitations of tra tional game-theoretic analysis for disaster management since decision-makers must anal dynamic and time-contingent strategies, preferences, and opportunities. In contrast, D constitutes a theory of mutual interactive game design in which decision-makers, by co municating with others prior to managing the disaster, interact to design the game that t will ultimately play (including players, options, and beliefs).

Finally, Strickert, Smarasinghe, Doscher, and Davies discuss policy development mountainous communities in New Zealand. They present a conceptual 'gap haz analysis' framework and methodology for policy development research in compl uncertain, and data-poor environments with the objective of sustainable hazard mitigation. Their research focuses on managing New Zealand 'ski fields' in the face of natural hazards and managing transient populations during emergency response. Their methodology integrates geomorphic assessments, qualitative interviews, and fuzzy cognitive maps with three stakeholder groups for simulating policy scenarios. It is shown how an integrated gap hazard analysis can be successfully embedded within a policy framework for increasing resilience to earthquakes in New Zealand ski areas. Key attention is given to policies that increase resilience to earthquakes or other low-probability, high-consequence events.

These five papers provide key insights into a new paradigm for disaster management: DRR and disaster resilience (ISDR, 2004). These emerging and holistic concepts are described in Table 1, as compared with traditional approaches to crisis and emergency management.

The principles of disaster risk reduction and disaster resilience can help communities to overcome pathological behaviour (including unsustainable livelihoods, institutional

Table 1. The evolution of risk assessment and emergency management as reflected in the five special issue papers

Crisis and Emergency Management Paradigm		Disaster Risk Reduction and Disaster Resilience Paradigm
1. Hazards, emergency and disaster focused		Vulnerability, risk and resilience focused (Balagh & Buchenrieder)
2. Single, event-based scenarios	**Emphasis**	Dynamic, multiple risk issues, and development scenarios (Levy)
4. Reactive: Respond to and recover from event	→	Proactive: assess, monitor, and update/adapt (Kousky; Alexander)
5. Fixed, location-specific conditions		Extended or changing conditions, with local variations (Levy)
6. Single authority or agency has responsibility	**Operations**	Involves multiple authorities and a multi-disciplinary approach (Alexander; Kousky; Levy)
7. Established hierarchical relationships: command and control	→	Shifting and fluid relationships: distributed decision-making, flexibility, and agility (Alexander)
8. Response and recovery		Mitigation and preparedness
9. Urgent, immediate and short time frames in outlook and planning	**Time Horizons** →	Moderate and long time frames in outlook and planning (Balagh & Buchenrieder; Strickert, Smarasinghe, Doscher, & Davies)
10. Rapidly changing, dynamic information usage		Accumulated, historical, layered, or comparative use of information (Levy; Alexander)
11. Communicating to communities	**Information Use and Management**	Communicating with communities (Levy; Strickert, Smarasinghe, Doscher, & Davies)
12. Directed, 'need to know' basis of information dissemination	→	Public information, diverse sources of information, and differing perspectives (Kousky)

Source: Modified from ISDR (2004).

rigidity, social dependencies, political hegemony, and ecological degradation) by vi ing human culture as embedded within nature; dependent on it, and capable of harm it. Specifically, this special issue has highlighted that lives and property can be secu through local sustainable disaster pre-planning, just as homes raised above the expec flood elevation ('homes on stilts') are better designed than many non-elevated structure absorb the shock of an inundation event. However, despite decades of conceptual deba there remain challenges related to the definition and operationalization of disaster risk resilience, particularly in the context of mitigating and adapting to the impacts of glo climate variability and change (as well as other global hazards and perils).

It is hoped that this special issue will further buttress the concept of disaster risk red tion as a key principle of disaster prevention, with the aim of reducing losses due to natu hazards, technologic disasters, and health-related crises. Recognizing that disasters threa not only ecosystems and economies but also human lives and cultural heritage, this s cial issue identifies three overarching principles for reducing disaster risk and enhanc community resilience. First, the use of scientific knowledge and leading-edge technol must be increased. This typically requires multidisciplinary and intersectoral partnersh among social scientists, engineers, and politicians. For example, the current generatio remote-sensing satellites provides reliable, detailed, cost-effective, continuous, real-ti and synoptic coverage of natural and anthropogenic phenomena (while the 'sensor w constitutes a distributed global web of sensors that can be used for sharing disaster infor tion among networked platforms). Second, successful disaster prevention policies req building a culture of disaster resilience. This involves creating grass-roots disaster e cation programmes and increasing public awareness about hazards and risk. Finally, i imperative that public officials marshal the political courage to implement disaster reduction policies: government officials must make significant investments in disaster r reduction in collaboration with regional organizations, non-governmental organizatio United Nations entities, and civil society representatives.

In summary, this special issue provides a collection of creative and effective to policies, and methodologies to help policy-makers prevent, prepare for, respond to, a recover from, natural, man-made, and health-related disasters. The devastation caused the BP Gulf Coast oil spill shows that traditional plans and policies for reducing dis ter risk are inadequate and that decision-makers must redouble their efforts to deve disaster-resilient communities and a culture of preparedness. The five papers in this spec collection provide original disaster risk-reduction strategies for coping with external sho and rebuilding when necessary—hence hazards and disasters are viewed as an opportur to build back stronger.

Finally, we would like to express our appreciation to the eight authors of the five pap and to the anonymous referees whose careful and timely reviews helped to sharpen ideas found in them.

References

Alexander, D. E. (1993) *Natural Disasters* (London and New York: UCL Press and Chapman & Hall).

BBC News (2010) Haiti quake death toll rises to 230,000. Available at http://news.bbc.co.uk/2/hi/americ 8507531.stm (accessed 10 February 2010).

Carter, M. R., Little, P. D., Mogues, T., & Negatu, W. (2007) Poverty traps and natural disasters in Ethiopia Honduras, *World Development*, 35(5), pp. 835–856.

Duncan, K., & Brebbia, C. A. (Eds) (2009) *Disaster Management and Human Health Risk: Reducing Risk, Improving Outcomes (WIT Transactions on the Built Environment)* (Southampton, UK/Boston, USA: WIT Press).

Holzmann, R., Sherburne-Benz, L., & Telsuic, E. (2003) *Social Risk Management. The World Bank's Approach to Social Protection in a Globalized World* (Washington, DC: Social Protection Department, The World Bank).

IFPRI (2002) Background information for presenters in improving understanding of risks and household responses to risk. Paper presented at the World Bank Conference on 'Risk and Vulnerability: Estimation and Policy Implications', Washington, DC, IFPRI, 23–24 September.

Inter-Agency Secretariat of the International Strategy for Disaster Reduction (UN/ISDR) (2004) *Living with Risk: A Global Review of Disaster Reduction Initiatives* (United Nations Publications, International Strategy for Disaster Reduction).

Mileti, D. (1999) *Disasters by Design: A Reassessment of Natural Hazards in the United States* (Washington, DC: Joseph Henry Press).

Potter, S. (2008) *The Sting of Climate Change* (Policy Brief, Sydney: Lowy Institute).

Prince, S. H. (1920) Catastrophe and social change: Based upon a sociological study of the Halifax Disaster, *Studies in History, Economics, and Public Law*, 94, pp. 1–152.

Sricharoen, T. (2006) Vulnerability and risk management for sustainable livelihoods of farm households in northern Thailand—The role of health insurance in managing risk, PhD thesis, Hohenheim University, 2006.

Quarantelli, E. L. (1998) *What Is a Disaster?* (London: Routledge).

Vos, F., Rodriguez, J., Below, R., & Guha-Sapir, D. (2010) *Annual Disaster Statistical Review 2009: The Numbers and Trends* (Brussels: CRED).

Wenger, D. (1985) Collective behavior and disaster research, in: R. Dynes, B. De Marchi, & C. Pelanda (Eds) *Sociology of Disaster: Contributions of Sociology to Disaster Research*, pp. 213–238 (Milan, Italy: Franco Angeli).

White, G. F. (1942) *Human Adjustment to Floods*, Research Paper No. 29 (Chicago: University of Chicago, Department of Geography).

Wisner, B., O'Keefe, P., & Westgate, K. (1976) Poverty and disaster, *New Society*, 37, pp. 546–548.

Wright, J. D., & Rossi, P. H. (1981) *Social Science and Natural Hazards* (Cambridge, MA: Abt Books).

Zimmerman, F. J., & Carter, M. R. (2003) Asset smoothing, consumption smoothing and the reproduction of inequality under risk and subsistence constraints, *Journal of Development Economics*, 71, pp. 233–260.

The L'Aquila Earthquake of 6 April 2009 and Italian Government Policy on Disaster Response

DAVID E. ALEXANDER
CESPRO—Centre for Risk and Civil Protection Studies, University of Florence, Florence, Italy

ABSTRACT *This paper describes the impact of the earthquake that struck the central Italian city of L'Aquila on 6 April 2009, killing 308 people and leaving 67 500 homeless. The pre-impact, emergency, and early recovery phases are discussed in terms of the nature and effectiveness of government policy. Disaster risk reduction (DRR) in Italy is evaluated in relation to the structure of civil protection and changes wrought by both the L'Aquila disaster and public scandals connected with the misappropriation of funds. Six of the most important lessons are derived from this analysis and related to DRR needs both in Italy and elsewhere in the world.*

Introduction: Lessons Learned?

It is common for books and articles about natural disasters to include the phrase 'lessons learned' in their titles (e.g. Yin & Moore, 1985; Potter, 2007). It is evident that there are still many lessons to learn and, judging by the frequent use of the phrase, there is also a certain willingness to learn them. However, *caveat lector*, as the natural hazards field is replete with lessons ignored, archived, forgotten, and generally not learned. There is a tendency both to repeat past mistakes and 'reinvent the wheel' by trying to innovate in ways that have already been well tried elsewhere. The test of lessons learned is that they be utilized and incorporated into current practice—in short that they have some influence upon the solution to the problem at hand (Berkes, 2007).

In the field of disaster risk reduction there has been a gradual realization that democratic participation in decision-making is essential to success; in other words that the key to risk and disaster management is governance (Özerdem & Jacoby, 2006). As Handmer and Dovers (2007, p. 30) noted in a seminal book on this topic, 'Policies emerge through complex and variable *policy processes* that include both government and non-government players'. Lessons thus need to be learned by a wide variety of interested parties, and this requires that they be disseminated, shared, and discussed.

This article addresses the disaster that struck the city of L'Aquila, central Italy, on 6 April 2009. It looks at government policy on earthquake prediction, emergency response, and the short- to medium-term management of the aftermath. For each phase, I will consider the principal lessons of the event and assess whether the Italian Government and its civil society partners made sufficient attempts to learn them, and to incorporate the lessons in policy.

The L'Aquila Earthquake

In October 2008 a sequence, or swarm, of minor earthquakes began in the central Apennine mountains east and southeast of Rome (Pacor *et al.*, 2009). It lasted until the following summer and included several thousand measurable tremors. Three of these were shallow focus events that were stronger than magnitude 5.0, and the largest of them reached magnitude $M_W = 6.3$. It occurred at 03.32 local time on Monday 6 April 2009 and had an epicentre 3.4 km southwest of the city of L'Aquila (population 72 800). L'Aquila is the capital of the province of the same name and of Abruzzo, one of Italy's 20 regions.

Three hundred and eight people were killed in the main shock and about 1500 were injured, 202 of them seriously (Volpini, 2009). Sixty thousand buildings were seriously damaged, including San Salvatore, the regional hospital and main trauma centre (Casarotti *et al.*, 2009). In total, 67 500 people were left homeless. Roughly a third were temporarily rehoused in hotels on the Adriatic Sea coast of Abruzzo and about a third were accommodated in 171 tent camps. Most of the others found alternative accommodation on their own initiative.

L'Aquila has a millennial history and a very substantial concentration of historic buildings in its centre. The entire historical quarter of the city was cordoned off and access was only allowed to bona fide individuals and groups accompanied by firemen and other officials detailed with the task of manning the barricades. Residents of the inner city could only visit their homes for very brief periods, and strictly under fire service escort. Besides the 14 outlying villages of L'Aquila municipality, another 15 towns registered severe effects and 31 were moderately damaged. Among the worst affected settlements, Onna, Paganica, San Gregorio, Tempéra, and Villa Sant'Angelo had their historical centres closed to the public and presided over by the police or fire services (Figure 1). The first of these, Onna, was a village of 300 inhabitants located on soft alluvial sediments in the valley of the River Aterno. Two thirds of the houses were completely ruined and 40 inhabitants died.

The L'Aquila earthquake of 6 April 2009 was a classic example of a medium-power seismic event. However, given the high vulnerability of building stock in the mountains of Abruzzo, it had a disproportionately large human impact. Had the tremors occurred, say, eight hours later, the death toll would probably have been several times larger, as the centre of the city would have been packed with visitors. L'Aquila is a university town and 66 of the 308 deaths were students in the age group 20–29 years, nine of whom died in the collapse of one wing of a university dormitory. The university is one of the principal sources of employment in L'Aquila and most of its buildings were severely damaged. Moreover, as they were situated in the city centre, many of them were also off-limits after the disaster. For some months, university functions, including classes, dormitories, and canteens, were transferred to tents.

The L'Aquila earthquake has already been extensively written about (e.g. Calvi *et al.*, 2009; Papanikolaou *et al.*, 2009; Rossetto *et al.*, 2009), but there has been a certain reluctance to consider the political and policy implications of the disaster from an academic point of view. In a popular vein, the Italian film director Sabina Guzzanti has

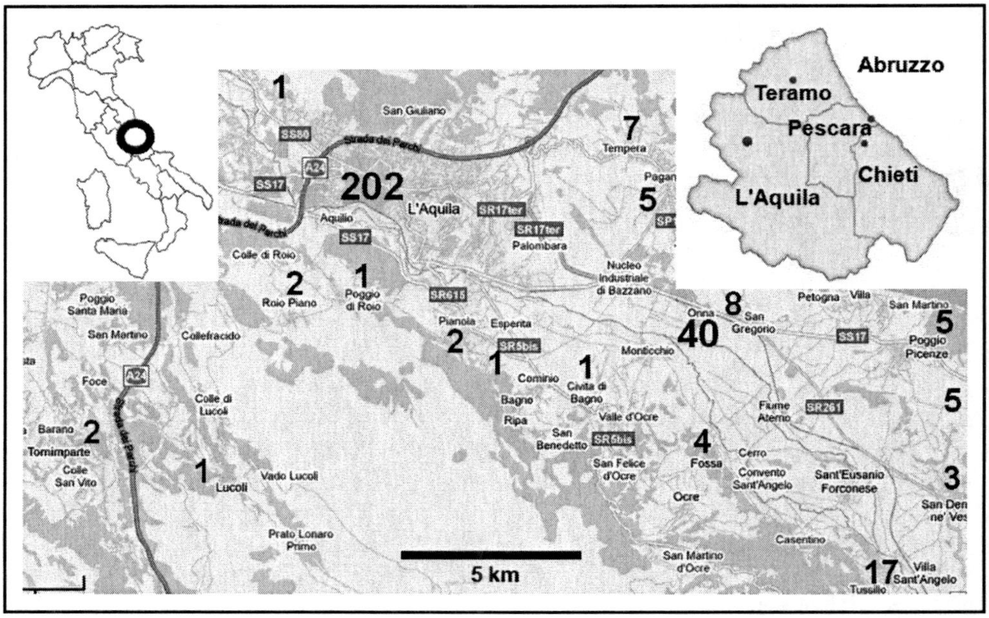

Figure 1. L'Aquila and earthquake area: location map and distribution of deaths.

made a documentary called *Draquila: Italy Shakes* (utilizing a conflation of 'Aquila' and 'Dracula') about the corruption and political incompetence that distinguished the aftermath of the disaster. It was feted at the Cannes film festival and has been very widely viewed in Italy. However, as the film was specifically intended to make a political point, a more sober academic analysis of government policy has yet to appear.

The calamity in Abruzzo is a good example of a moderate earthquake, albeit with immoderate effects, and these deserve more attention by students of disaster, who tend to devote most of their energies to studying the larger events such as those in Pakistan in October 2005 and Haiti in January 2010. Moderate earthquakes are more common than the larger events, and, as the L'Aquila example shows, can cause disproportionate suffering. They therefore need to be studied more intensively, especially from a policy point of view.

The L'Aquila disaster was the first earthquake to have occurred in Italy with its epicentre very near to a major population centre since the Avezzano catastrophe of 1915, which killed 34 000 people, including 94% of the population of Avezzano itself, a town barely 100 km from L'Aquila. Moreover, the 1908 Strait of Messina earthquake and tsunami killed an estimated 128 000 people on the peninsular mainland and in Sicily. Hence, although the L'Aquila disaster was only a moderate seismic event, it could be considered as a rehearsal or test of the Italian Government's ability to manage a much larger seismic calamity, which, unfortunately, cannot be ruled out in the future.

The following sections will consider policy before, during, and after the earthquake emergency, covering a period that extends from early 2009 to mid-2010.

The Structure of Disaster Management in Italy

Response to the threat and occurrence of major emergencies in Italy is divided between the country's civil defence and civil protection systems. The former is the means of combating terrorism and is centralized under the command of the Minister of the Interior in the national government. In Italy it is articulated in the provinces through the country's 109 prefects, who are the chief provincial civil servants and leading representatives of the central state at the intermediate level of government. Preparation for chemical, biological, radiological and nuclear (CBRN) terrorist incidents is in the hands of the National Fire and Rescue Service, which maintains centres of expertise in this field in major cities such as Milan, Bologna, Rome, and Naples. The state police force, Italian armed forces, and SISDe, the Secret Service, cover most of the other tasks involved in counter-terrorism (Palmieri, 2003).

Whereas civil defence enjoys a 'top-down' organization, civil protection is much more 'bottom-up' (Alexander, 2002a). It nevertheless has a national articulation that has evolved over several decades between one disaster and another. A law of 1970, no. 996, organized emergency response around the National Fire and Rescue Service, on the grounds that in major emergencies such as large earthquakes the first requirement is usually urban search and rescue coupled with the stabilization of the site, which the Fire Service is uniquely qualified to achieve. It thus became the 'lead agency' in civil protection. In response to a series of disasters, particularly the Irpinia (central-southern Italy) earthquake of November 1980, which killed 3000 people and injured 8800, in 1982 civil protection was established as a ministry without portfolio under the aegis of the Ministry of the Interior. This followed years in which the response to emergencies was spontaneous and led by a government-appointed commissar, whose role had never fully been institutionalized (Alexander, 2002b).

It took 10 years to establish the legal basis of a full national structure for civil protection, which finally emerged in Law no. 225 of 1992. Interestingly, in his role as guarantor of the Italian Constitution, President Francesco Cossiga refused to sign the 1991 draft of this law on the grounds that the Constitution made no provision for emergency management and thus could not guarantee that the law would not infringe the rights and freedoms of the Italian people. This observation remains highly significant and will be re-examined in the light of the response to the L'Aquila earthquake.

In 1999–2000 the Ministry of Civil Protection was briefly disestablished and converted into an independent agency, a form of QUANGO (quasi-autonomous non-governmental organization). However, after three months a new model was adopted which made civil protection a department of state, with nominal dependence on the Ministry of the Interior but direct responsibility to the Prime Minister as head of the national Cabinet. The director of civil protection had the status of an undersecretary of state (Alexander, 2002b). In this form national civil protection survived for almost a decade. Indeed, this model of organization became a non-binding directive of the European Union, and was adopted by, for example, the United Kingdom and Sweden. It makes sense, as civil protection is a transverse responsibility that extends across various ministerial competencies, such as health, public works, economy, and employment.

Meanwhile, civil protection structures grew solidly at all the other levels of public administration. The 1992 law organized these into a coherent system, which included as its backbone the country's 3600 emergency response volunteer organizations, some of which have existed for as long as seven centuries. A legislative decree of 1998 decentralized public administration and two rather vaguely formulated articles transferred civil protection from the provincial prefects (again, representatives of the central state) to the regional and provincial governments, which are elected by local populations. It is important to note that the 1992 law, no. 225, unequivocally places the elected mayors of Italy's 8104 municipalities as the executive heads of civil protection. Technically, if not practically, in dealing with local emergencies they have more power than the Prime Minister, and it is as well to remember that all disasters are local affairs, as the local area is always the theatre of response operations (Alexander, 2002c).

Italian civil protection is organized on a cascading basis, with linked emergency operations centres (EOCs) that extend from the centre of national government to the smallest local level: *DICOMAC*—national command and control EOC; regional government EOC; *SOUP*—provincial and prefectural EOC; *COM*—mixed, or nodal, municipal operations centre; and *COC*—local municipal operations centre. According to the 'Augustus System' (named after the Roman Emperor in view of his opinions on command and control), communication between the levels, and between organizations, is maintained on a sectoral basis using an adapted version of the US Federal Emergency Management Agency's emergency support functions (ESFs). These cover 15 sectors, including transportation, communications, logistics management, and search and rescue (Stallings & Faust, 2009). In the USA, the ESFs help coordinate support in an incident or disaster among the numerous federal agencies at work. However, in Italy the support functions extend to all levels of public administration, from national to local.

A recent report by the Organization for Economic Co-operation and Development (OECD, 2010) concludes that the Italian model of civil protection is basically sound and contains much of merit, but it is very much a mosaic of varied levels of development in which there is much more that needs to be achieved. As the authors of the report concentrated on investigating the official protagonists of civil protection, especially members of the national Department of Civil Protection (DPC), and as they relied on self-assessments by these protagonists, they failed to deal adequately with many of the independent criticisms of the system. These will be summarized later in this article. However, I now consider the response of the system to the country's first major seismic emergency since 1997.

The Pre-Earthquake Phase

Mr Gioacchino Giampaolo Giuliani is a technician who works in the national physics laboratory located in tunnels under Gran Sasso, peninsular Italy's highest mountain, which is 40 km east of L'Aquila. For some years his hobby has been to monitor atmospheric radon emissions in an attempt to correlate them with heightened seismic activity and thus gain a means of making short-term predictions of violent earthquakes. He maintained a radon meter in the basement of a school in L'Aquila and took frequent readings from it. He noted increases in radon emissions shortly before the October 2002 San Giuliano di Puglia earthquake, which occurred 240 km southeast of L'Aquila but was unable to pinpoint the epicentre.

Radon is an inert gas that is transferred in increasing quantities to air or groundwater when micro-fracturing around an active geological fault is induced by seismic strain. Radon emissions have been used to map the structure of active faults (Ciotoli *et al.*, 1998), including in the Abruzzo Apennines (Bella *et al.*, 2001). However, it should be emphasized that with current knowledge and technology, earthquakes cannot be predicted in the short-term with such measurements. The L'Aquila earthquake was preceded by anomalies in radio waves (Biagi *et al.*, 2009), earthquake lights (Fidani, 2010), magnetic irregularities (Masci *et al.*, 2009), seismic foreshocks (Papadopoulos *et al.*, 2010), uranium fluxes in groundwater (Plastino *et al.*, 2009), and ionospheric changes (Tsolis & Xenos, 2010). However, none of these precursors would have enabled a short-term prediction to be issued to the public.

After a few days in which he vacillated when questioned by journalists, on 29 March 2009 Mr Giuliani informed the mayor of the town of Sulmona (population 25 200) that a 'catastrophic' earthquake was about to happen. His prediction was followed within 24 hours by a seismic event of magnitude 3.8 that, of course, was too small to produce damage. Moreover, it was one of 900 tremors registered over the six months that preceded the L'Aquila main shock. However, word had got around and the population of Sulmona busied itself taking precautions, albeit not in a systematic way and without activating the town's emergency plan. According to the local newspaper *Il Centro*, priests hurried to remove relics from churches and people dragged bedding into the street. The Italian National Department of Civil Protection (DPC) began legal proceeding and asked for punitive damages because Giuliani was alleged to have created unnecessary alarmism. Moreover, the National Astrophysics Institute was careful to point out that Mr Giuliani's predictions were not made in the context of his job and were not sanctioned by his employer.

As a result of the ensuing chaos, at 18:30 hrs on Tuesday 31 March 2009 the DPC convened the national Commission on Major Risks (*Commissione sui Grandi Rischi*) in L'Aquila. Present at the meeting were more than a dozen officials from the national Department of Civil Protection, universities, and regional and local government. The minutes were later published by a weekly magazine (*L'Espresso*, 2009). The two-page summary of the one-hour meeting shows considerable complacency, some oversimplification, and a degree of evasiveness. Its main conclusions were articulated by Professor Franco Barberi, a volcanologist at the University of Rome III and former Undersecretary of State for Civil Protection:

> Professor Barberi concluded that there was no reason to say that a sequence of low-magnitude shocks could be considered as the precursor of a major event.
>
> Referring to what was happening at L'Aquila, Professor Barberi explained that the measurement of radon gas in order to predict earthquakes is an old problem that has long been studied without obtaining useful solutions. (L'Espresso, 2009)

The main shock occurred one week after Mr Giuliani had predicted it, and with an epicentre located just outside the city of L'Aquila, 55 km away from Sulmona.

In a television interview after the disaster, Mr Giuliani stated that radon measurements had reached such a peak on the night of 5 April 2009 that three hours before the main shock he made his family leave their home and spend the night outside. However, during another television appearance after the disaster he contradicted this affirmation and said that they had remained at home. The televised debates, which went on for more than a week, were vitriolic and involved a welter of accusations and counter-accusations. Giuliani demanded that the authorities beg his pardon, but they were disinclined to do so.

My own position, as explained to the European mass media (and widely quoted on Italian television (RAI News 24, 2009)) immediately after the disaster, was that whether or not a prediction was possible was a red herring: unless the organizational means to deal with it existed, including adequate deployment of resources, little could be done to manage the emergency. Civil protection in Abruzzo was undeniably poorly developed and distinctly under-resourced. In short, it was unable to cope in any way with a major disaster.

After decades of research by seismologists all over the world, short-term earthquake prediction remains an elusive goal and a controversial topic (Holliday *et al.*, 2005). Limited successes unfortunately do not add up to a reliable means of identifying the window of time, in hours, in which an earthquake may occur. However, in the central Apennines the more pressing scientific question is whether earthquake sequences are made up exclusively of low-level tremors or whether they can include a devastating tremor. In this context, the 1997 swarm, which occurred further north in Umbria and Marche regions, included three events of magnitude in the range 5.4–5.6 (Zollo *et al.*, 1999). However, the majority of the damage was essentially the result of the cumulative effects of numerous earthquakes spread over a three-month period, and perhaps that is also true of the 13 fatalities. Nevertheless, the L'Aquila earthquake swarm of 1703 involved a main shock that is reputed to have killed 10 000 people.

As articulated at the meeting of the Commission on Major Risks on 31 March 2009, the Italian Government's position was unequivocal: there was no cause for alarm. This attitude permeated its way down the ranks of the civil protection system. Then, at 00:30 hrs on Monday 6 April 2010, a tremor that was larger than usual shook L'Aquila. Residents rushed out of their houses in alarm. The strategy adopted by civil protection authorities was to tour the streets with loudspeakers advising people to calm down and return home. In the town of Pagánica, less than 10 km northeast of L'Aquila, residents did exactly that: in the ensuing main shock three hours later, eight of them died and 40 were seriously injured. In L'Aquila city I investigated one case in which a young lady had decided to remain out of doors after the foreshock, while her parents returned home. Their bodies were recovered by firemen from a space barely 15 cm wide into which the building had compressed as it collapsed (Figure 2).

These observations now require a more general description of the immediate post-impact phase and how management of the emergency was affected by government policy on civil protection.

The Emergency Phase

L'Aquila is 117 km, or about 90 minutes' drive, from the centre of Rome via a good *autostrada* expressway, which suffered little damage in the earthquake. Response to the disaster was rapid and involved the same concept of 'overwhelming force' that was so effective in subduing Saddam Hussein in Iraq during the Allied invasion of 2003. Hence, the Italian Government's policy on the issue of immediate disaster response was to ensure it went well by encouraging an indiscriminate convergence reaction involving massive amounts of resources. This was perhaps as well, given that the main trauma hospital in the area had suffered major structural damage (concentrated, unfortunately, on the Accident & Emergency reception area). Within 24 hours it was completely replaced by the first of two field hospitals that arrived in the area. Meanwhile, early management of casualties was largely in the hands of the Italian armed forces, which employed 'Medivac' aerial evacuation techniques to remove seriously injured patients to hospitals outside the disaster area.

Figure 2. Mid-floor compression of a multi-storey reinforced concrete apartment block in L'Aquila city leading to the deaths of two inhabitants.

As luck would have it, at Coppito just outside the city of L'Aquila lies the national training centre for the country's financial police force (*Guardia di Finanza*), which has a spacious, secure compound, and plenty of accommodation, including residential units. It was an ideal site from which to manage the emergency. The national on-site emergency operations centre (the DICOMAC—see above) was rapidly established in the sports hall of this facility and the managers of 16 support functions began to coordinate their activities. Volunteer organizations, the Italian Red Cross, and the armed forces also set up bases in this hall, which functioned as the nerve centre of emergency management for some months.

The earliest actions included the total evacuation of town centres, including the entire historical core of L'Aquila, the first time in modern Italian history that a major city had been completely emptied of its population by government decree. They also involved setting up seven districts based on municipal mixed operation centres (COMs—see above) with links to local EOCs in all 57 municipalities affected by the disaster. One of the principal tasks of these was to set up and manage the 171 tent camps that were established for homeless survivors and provide safe, secure environments for them. Two of the districts were managed by military commands and here the local populations protested that the military had overreached themselves in maintaining security to the extent that the rigid regulations and tall fences around the camps made them seem like prisons.

Under the guidance of the Prime Minister, Mr Silvio Berlusconi, and his Cabinet, the emergency management policy was unequivocally one of replacing the local response with a national one. For example, the legal framework clearly indicated that the decision to evacuate L'Aquila city centre fell to its mayor, Mr Massimo Cialente, a 55-year-old former hospital doctor of the left-wing Democratic Party. It was, instead, brusquely taken over by the national Department of Civil Protection (DPC), and so were many other local functions. None of this did anything to improve local resiliency or self-sufficiency, which effectively disappeared. Nevertheless, the management of the emergency was undoubtedly a success, if only because of the overwhelming national response. Three weeks after the disaster the majority of traffic on the local roads consisted of emergency vehicles from the police, fire and medical services, and armed forces, and that situation persisted for months.

Paradoxically, the Italian civil protection system is remarkably non-hierarchical. The adoption of support functions for the sectoral management of resources and abundant use of information and communications technology obviate the need for complex chains of command, which in any case would not be tolerated by the volunteer organizations—the backbone of the system—which are happy to collaborate but answer only to their own command structures. In some respects this is a very advanced form of civil protection, a field in which demilitarization and movement towards the collaborative model are generally signs of progress (Alexander, 2008). However, in Italy its Achilles' heel is the failure to complete the system by fully articulating the means of planning for emergencies, managing them and dealing with disaster risk at all levels of government and society. The national guidelines for these processes are excessively complex, out of date and out of step with the current ethos of federalism. One can hardly escape the conclusion that not only has the Italian government's lack of interest in disaster management caused this predicament, but its desire to leave the system incomplete is a convenient means of retaining control. This situation shows how strong can be the tension between federalism—the desire to gain votes by ceding control to local administrations—and centralism—the desire to retain control by not ceding any such power (Handmer & Dovers, 2007, pp. 64–76).

The Intermediate Term

Earthquake disasters commonly lead to massive demand for alternative housing, and correspondingly massive needs for structural survey of existing building stock to determine whether it is serviceable, capable of being rehabilitated, or needing to be demolished (Choudhury & Jones, 1996). It is usual, perhaps even normal, for there to be a sequence of shelter, which begins with emergency accommodation, probably in tents, proceeds to transitional shelter, in container homes or prefabricated buildings, and ends with full reconstruction, which may take more than 10 years if the damage is extensive and complex (Figure 3, see also Lizarralde *et al.*, 2010).

In the wake of the L'Aquila earthquake, the Italian Government's policy on shelter was partially to avoid the intermediate stage. Although regional governments and charitable foundations donated prefabs, the main thrust of central government policy was to keep people in tents for up to seven months and rehouse them in complexes of buildings that are too substantial to be regarded as transitional, but too unusual to be described as permanent rehousing (Figure 4). A total of €1.6 billion ($1.4 billion) was spent on constructing 5736 apartments in 19 complexes (Figure 5), the CASE project (*Complessi Antisismici Sostenibili ed Ecocompatibili*—Calvi & Spaziante, 2009). Antiseismic design was embodied in the expensive mechanisms of base isolation underpinning some of the complexes. Although designed with modern environmental compatibility in mind, some of the complexes are not connected to wastewater treatment and discharge raw sewage directly into

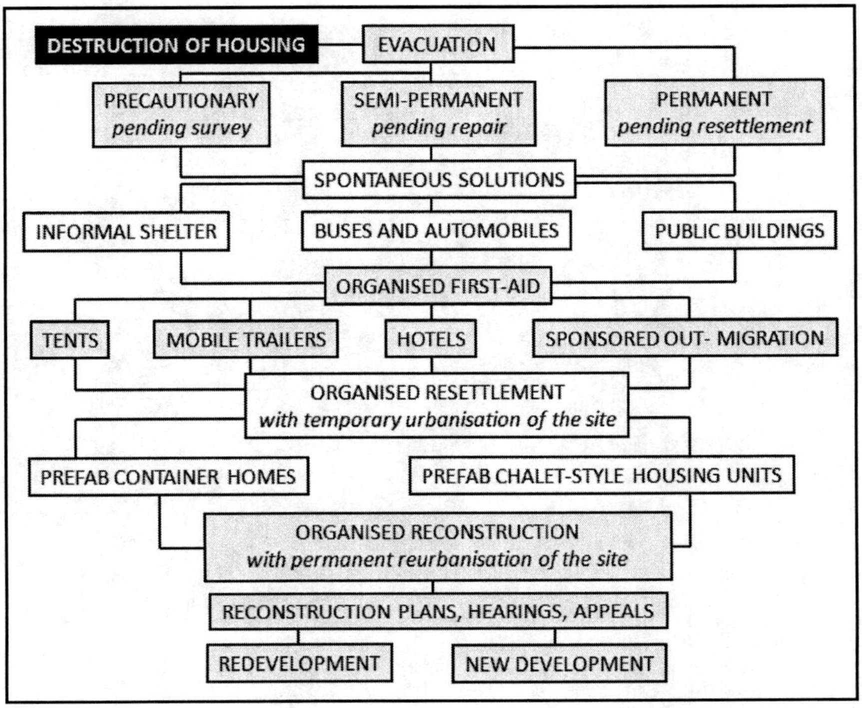

Figure 3. Full sequence of post-disaster shelter.

the Aterno River. Moreover, all were established on greenfield sites and some were built in pre-existing environmental conservation areas. Nonetheless, the CASE project was completed in six months flat, including landscaping, primary urbanization, and access roads: it had to be, as the Aquilan winter is too cold and harsh for survivors to be left without proper accommodation.

To achieve the direct transition from homelessness to secure accommodation in the CASE units, survivors remained in tents (22 000 people) and coastal tourist accommodation (21 000 people) throughout the summer. From beginning to end, there were psychological problems associated with the fragmentation of pre-existing communities, especially among children.

Central government policy on rehousing the survivors concentrated heavily, indeed almost exclusively, on the CASE project. Judged in its own terms it was a distinct success. Residents were provided by the state with all basic necessities, right down to cutlery and bed linen. However, the complexes lack virtually all external services. Most of them are detached, even distant, from shops and other amenities. The infrastructure of the Aterno Valley is completely inadequate and has not been upgraded. Moreover, nothing has been done to stimulate employment. For lack of accommodation, service industries and commercial concerns have moved out of the area, and it is estimated that 26 000 people have been made unemployed by the earthquake—in an area in which the unemployment rate, at 12.2%, is higher than the national average and youth unemployment is officially 40%, although in practice much higher.

Figure 4. (Above) CASE post-earthquake housing at Bazzano on the periphery of L'Aquila city. (Below) Detail of antiseismic base isolation on the CASE units.

The hardships of relocation were exacerbated by financial burdens. Many people were forced to continue paying local taxes and mortgages on properties that had been destroyed. The Government put a moratorium on taxes until 2010 but then demanded that people

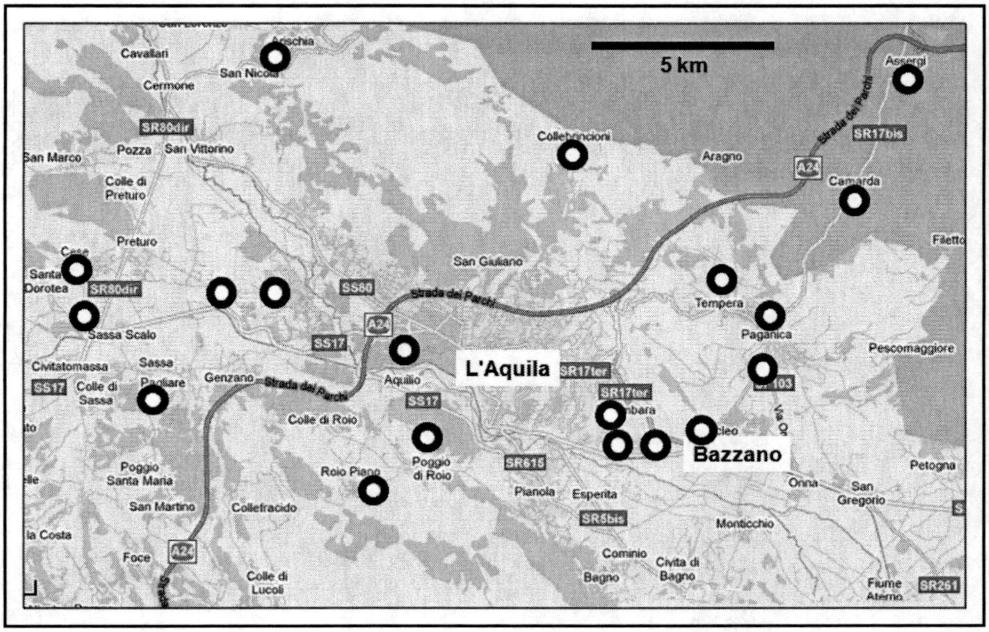

Figure 5. Location of the CASE post-seismic housing complexes in the L'Aquila area.

repay all taxes foregone during the moratorium period. After public outcry the repayment period was extended from five to 10 years. Nevertheless, it is hardly surprising that the local economy has stagnated. It had little to stimulate it before the earthquake.

A year after the disaster the historic buildings at the centre of the towns and L'Aquila city remained off limits. The Government had organized a major programme of buttressing, on a scale and of a sophistication never seen before. Clearly, the wood, scaffolding, and electro-soldered steel buttresses were designed to last for a very long time (Figure 6). At the same time, the Government's emergency managers neglected to deal with the problem of the four million tonnes of rubble cluttering the streets of the interdicted areas. They belatedly got to grips with this when the population organized a mass protest that broke through the barriers and symbolically carried away lumps of debris and wheelbarrows full of rubble.

The missing element in the Italian Government's recovery policy is local participation. It has been replaced by a form of state paternalism that is totally out of kilter with the present worldwide mood of neoliberalism. The reason lies in vote garnering. In the regional and provincial elections of June 2009, Prime Minister Berlusconi's Freedom-People Party won control of all four provinces and the Region of Abruzzo. This was achieved on the basis of populist measures such as the promise of free housing for all survivors, and consolidated through some 29 prime ministerial visits to the area. The cost of these measures was enormous. Their benefit to the recovery process was highly debatable, in that it is well known that the more substantial, integral, and comfortable one makes the temporary housing, the less incentive there is to reconstruct permanent accommodation. Moreover, the neglect of the economy and infrastructure failed to kick-start any indigenous form of recovery.

Figure 6. Buttressing of a severely damaged building in the historical centre of L'Aquila city.

Fifteen months after the disaster the situation was as follows. Government intervention coupled with more local charitable initiatives had succeeded in rehousing more than 90% of the homeless survivors in need of assistance, and the others would soon be accommodated. The local economy, employment, services, and demography were in decline. The participation of the local public had been stymied and discontent was rife. On 6 July 2010 the people of L'Aquila marched on Rome in a peaceful demonstration that was brutally broken up by state police, who even battered the Mayor of L'Aquila with truncheons. Central government ignored it. Numerous emergent groups had been created in the disaster area (for example, see www.3e32.com/), but none of them had found enough of a voice to represent the groundswell of discontent in the national arena.

Prognosis for the Long Term and Reconstruction

The quality of prefabricated temporary housing and associated urbanization, and the unresolved status of the CASE project complexes, point to a recovery of long duration. So does the extensive buttressing of ruined buildings in the historical town and city centres. It is as well to remember that vestiges of the temporary housing still exist in Messina and Avezzano following earthquakes in 1908 and 1915, respectively. Moreover, inhabited prefabs can still be found in the Belice Valley in western Sicily, the location of a 1968 earthquake. The upshot of this is that government policy is clearly angled towards the short term. Once people have been housed in accommodation that will last for years, they can be forgotten, or at least left to better their circumstances without further state intervention.

This is the new form of liberalism (Stacul, 2007) After the 1908 Strait of Messina earthquake the liberal Prime Minister, Giovanni Giolitti (1842–1928), defended his government's poor response to the disaster by saying that it was the responsibility of the citizen to protect himself against natural hazards and the government did not have, and could not ask the citizen for, the necessary resources to do it (Boatti, 2008, ch. 11). By the 1960s the principle of 'assistentialism' had become firmly fixed in the national consciousness—i.e. that the state would indemnify the citizen against disasters, whatever it cost. Hence the state's bill for the 1980 Irpinia earthquake was €30 billion ($39 billion), plus a four eurocents tax on gasoline, apparently in perpetuity (Negri, 1996).

Given the ballooning costs of disaster, the question was *when*, not whether, conservative neoliberalism would take action to reduce them. Unfortunately, it did not do so directly by establishing ground rules about how much and for what the state would pay. Instead, the first response was absolute open-handedness, in which the government paid up front for massive emergency relief and prompt resettlement. Unless the situation changes in currently unpredictable ways, the probable long-term response is stagnation and neglect, which are the only fiscally responsible ways of compensating for the previous largesse.

Before summarizing the lessons of this analysis, a brief explanation is needed concerning what impact the disaster has had on the organization of civil protection in Italy.

Developments in Civil Protection After the L'Aquila Earthquake

In Italian emergency response, political scandals seem to recur about once every decade. In fact, there have been three in the last 30 years, and each has opened the door to radical change, especially regarding the form of political control of the service. The background to the latest scandal is intriguing. In major disasters one of the most widely used measures is the ordinance, a government decree with the status of law. For example, a municipal mayor may order the evacuation of an area of housing threatened by hazards using an ordinance that makes the process mandatory. Ordinances can also be issued by government ministers, including the Prime Minister, and by provincial prefects.

In civil protection, ordinances are a quick and easy administrative means of solving problems. So convenient had they become that between October 2001 and May 2009 the head of the national Department of Civil Protection (DPC) had signed 587 emergency ordinances and €10.6 billion ($13.6 billion) had been disbursed by means of them, yet without the scrutiny of parliament. In December 2009 this situation was consolidated when the Italian Government issued a Decree-Law (DL no. 195 of 30 December 2009) in which Article 16, carefully buried amidst other matters, privatized the national civil protection service.

For six weeks this extraordinary move solicited next to no public reaction. Trade unions protested weakly when 200 of the 650 employees of the DPC were transferred to the new holding company, and so did some of the people in question. The implications of DL 195, Article 16, were unfathomable. A theoretical case for such a move can be found in Kirschenbaum (2003, ch. 8), but in this case the arguments against it had not been aired in public. In brief, they mainly involve equity in relief and assistance, access to aid, public scrutiny of accounts and sourcing of funds. Eventually, the scandal broke and public reaction was vituperative. The measure was hastily rescinded. The main basis of the attack by the mass media and prominent members of the public was as follows. The use of ordinances had got out of hand and large amounts of public money had fallen into the hands of a group of unscrupulous engineers and entrepreneurs. The transformation of the source of these ordinances into a private company would accelerate that process and further isolate it from the scrutiny of parliament. Suddenly, Italy seemed to have stopped sleepwalking backwards towards the rampant corruption of the 'first republic' (Guzzini, 1995).

At the time of writing it is unclear what implications the scandal will have for the future organization of civil protection. There is little doubt that the 'troops on the ground', mainly the legions of volunteer relief workers, feel betrayed. However, in mid-2010 there is little sign of any attempt to restore credibility at the national level.

As a coda to this narrative, it is interesting to note that many buildings damaged in the April 2009 earthquake had not been erected in contravention of the prevailing seismic building codes. Until a comprehensive revision of 2006, the codes put L'Aquila in the moderately seismic category, rather than the high risk one (Romeo & Pugliese, 2000). Given the extensive damage and loss of life in the 1703 earthquake, this is hard to understand in scientific terms. However, it must have been very convenient to building firms, as it allowed cheaper and more lax standards of construction. The results of their good fortune can be seen in the smooth reinforcing bars, friable cement, poor joints, and lack of stiffness of reinforced concrete buildings that collapsed all over the area.

Lessons to Learn

As noted above, the L'Aquila earthquake of 6 April 2009 was a medium-power seismic event that caused a disproportionately large amount of damage and suffering. It revealed colossal amounts of vulnerability, which had been, and continues to be, poorly tackled by Italian governmental policy. As in other events, it seems to take a national disaster to bring problems to the attention of people and their elected representatives. For example, it took the deaths of 26 small children in the collapse of a school during the October 2002 Molise (central Italy) earthquake to create a programme for seismically retrofitting buildings in the Italian educational system (Grant *et al.*, 2007).

There follows six of the most important policy lessons of the L'Aquila disaster:

(1) Whether short-term earthquake predictions are official or amateur, reasonable or far-fetched, the precautionary principle should be employed, especially when local seismicity is already in an enhanced state. There should be no room for complacency and inaction. Similar challenges were encountered with the inadequacy of the official reaction to the Iben Browning earthquake prediction in southern Illinois in 1990 (Spence *et al.*, 1993).

(2) Scientific processes of monitoring and alert need to be backed by robust emergency response mechanisms with a local basis. Perhaps the most notorious violation of this principle in the last half century was the failure to warn local residents of the eruption of Nevado del Ruíz in Colombia in November 1985, leading to the deaths of 23 000 people (Voight, 1990).

(3) Local responses to catastrophe need to be augmented, not supplanted, with external assistance. This requires an integrated system of emergency response in which the barycentre is the local level, not the national one which should concentrate on harmonizing, co-ordinating and supporting local efforts. A similar need to implement this principle was evident during the emergency response to Hurricane Katrina (Bier, 2006).

(4) Safeguards against corruption and abuse of the system need to be instituted. Civil protection is in a unique position, in that it must necessarily have recourse to emergency

powers, but these need to have proper democratic underpinnings. Cooper and Block (2006) made the same case in minute detail in their political analysis of the US federal response to Hurricane Katrina.

(5) In most disaster response and emergency management systems there is a tension between centrism and devolution. This needs to be resolved with clarity and firmness, as each level of the hierarchy of public administration needs to know exactly what its responsibilities are. In the US context, Eisinger (2006) described such a situation as 'imperfect federalism'.

(6) The price of progress is in disaster risk reduction is eternal vigilance. Gains made in one year can be reversed in another when budgets are cut and programmes are reduced. That this should occur in the midst of the 'window of opportunity' created by recent events is unusual but not without precedent. For instance, when discussing the effect of the 1985 Mexico City earthquake on seismic retrofit policies in Costa Rica, Lavell (1994, p. 99) noted: 'And yet, despite all these factors, the Mexico City disaster per se did not induce the Caja authorities [the Costa Rican Social Security System] to initiate widespread vulnerability analyses'.

In conclusion, I have endeavoured to show that these lessons are universal. In Italy they have not yet been acted upon, but there is much scope for improvement. The 2010 OECD study concluded that Italian emergency response is basically sound and the system has some points of excellence (OECD, 2010). Given the magnitude of the hazards that the country faces, one can only hope that the neglected parts of it are brought up to scratch as soon as possible.

Acknowledgements

I would like to acknowledge the support of the European Commission's FP6 Project MICRODIS, administered by the Centre for Research on the Epidemiology of Disasters at the Catholic University of Louvain, Brussels, and of the Project's leaders Prof. Debarati Guha-Sapir and Ms Laura Irvine. I am also pleased to acknowledge the contribution of the MICRODIS-Aquila research team: Caterina Antinori, Anna Carbonelli, Vincenza Cofini, Diego Guidotti, Christian Iasio, Michele Magni, Fausto Marincioni, Roberto Miniati, and Marco Morabito.

References

Alexander, D. E. (2002a) From civil defense to civil protection—and back again, *Disaster Prevention and Management*, 11, pp. 209–213.

Alexander, D. E. (2002b) The evolution of civil protection in modern Italy, in: J. Dickie, J. Foot, & F. Snowden (Eds) *Disastro! Disasters in Italy since 1860: Culture, Politics, Society*, pp. 165–185 (New York: Palgrave Press).

Alexander, D. E. (2002c) *Principles of Emergency Planning and Management* (Harpenden, UK: Terra Publishing, & New York: Oxford University Press).

Alexander, D. E. (2008) Emergency command systems and major earthquake disasters, *Journal of Seismology and Earthquake Engineering*, 10, pp. 109–118.

Bella, F., Azzi, C., Bella, R., Della Monica, G., Plastino, W., Scandone, R., & Sgrigna, V. (2001) Tectonic deformation events and local seismicity in the Gran Sasso area of the central Apennines, *LNGS Annual Reports from 1997 to 2001*, pp. 169–178 (Gran Sasso d'Italia, Italy: Laboratorio Nazionale di Gran Sasso).

Berkes, F. (2007) Understanding uncertainty and reducing vulnerability: Lessons from resilience thinking, *Natural Hazards*, 41, pp. 283–295.

Biagi, P. F., Castellana, L., Maggipinto, T., Loiacono, D., Schiavulli, L., Ligonzo, T., Fiore, M., Suciu, E., & Ermini, A. (2009) A pre seismic radio anomaly revealed in the area where the Abruzzo earthquake (M = 6.3) occurred on 6 April 2009, *Natural Hazards and Earth Systems Sciences*, 9, pp. 1551–1556.

Bier, V. (2006) Hurricane Katrina as a bureaucratic nightmare, in: R. J. Daniels, D. F. Kettl, & H. Kunreuther (Eds) *On Risk and Disaster: Lessons from Hurricane Katrina*, pp. 243–254 (Philadelphia: University of Pennsylvania Press).

Boatti, G. (2008) *La Terra Trema: Messina, 28 Dicembre 1908. I Trenta Secondi che Cambiarono l'Italia, Non Gli Italiani* (Milan: Mondadori).

Calvi, G.M., & Spaziante, V. (2009) Reconstruction between temporary and definitive: The CASE project, *Progettazione Sismica*, 3, pp. 221–250.

Calvi, G. M., Stucchi, M., Meletti, C., Manfredi, G., & Dolce, M. (Eds) (2009) *Progettazione Sismica 03* (Pavia, Italy: IUSS Press, University of Pavia).

Casarotti, C., Pavese, A., & Peloso, S. (2009) The seismic response of the San Salvatore hospital of Coppito (L'Aquila) during the earthquake of April 6th, 2009, *Progettazione Sismica*, 3, pp. 159–172.

Choudhury, G. S., & Jones, N. P. (1996) Development and application of data collection forms for post-earthquake surveys of structural damage and human casualties, *Natural Hazards*, 13, pp. 17–38.

Ciotoli, G., Guerra, M., Lombardi, S., & Vittori, E. (1998) Soil gas survey for tracing seismogenic faults: A case study in the Fucino basin, central Italy, *Journal of Geophysical Research*, 103(B), pp. 781–794.

Cooper, C., & Block, R. (2006) *Disaster: Hurricane Katrina and the Failure of Homeland Security* (New York: Times Books).

Eisinger, P. (2006) Imperfect federalism: The intergovernmental partnership for homeland security, *Public Administration Review*, 66, pp. 537–545.

Fidani, C. (2010) The earthquake lights (EQL) of the 6 April 2009 Aquila earthquake in Central Italy, *Natural Hazards and Earth System Sciences*, 10, pp. 967–978.

Grant, D. N., Bommer, J. J., Pinho, R., Calvi, G. M., Goretti, A., & Meroni, F. (2007) A prioritization scheme for seismic intervention in school buildings in Italy, *Earthquake Spectra*, 23, pp. 291–314.

Guzzini, S. (1995) The 'long night of the first republic': Years of clientelistic implosion in Italy, *Review of International Political Economy*, 2, pp. 27–61.

Handmer, J., & Dovers, S. (2007) *Handbook of Disaster Policies and Institutions* (London: Earthscan).

Holliday, J. R, Nanjo, K. Z., Tiampo, K. F., Rundle, J. B., & Turcotte, D. L. (2005) Earthquake forecasting and its verification, *Nonlinear Processes in Geophysics*, 12, pp. 965–977.

Kirschenbaum, A. (2003) *Chaos, Organization, and Disaster Management* (New York: Marcel Dekker).

Lavell, A. (1994) Opening a policy window: The Costa Rican Hospital Retrofit and Seismic Insurance Programs 1986–1992, *International Journal of Mass Emergencies and Disasters*, 12, pp. 95–115.

L'Espresso (2009) Verbale della Commissione Grandi Rischi, L'Aquila, 31 Marzo 2009, 17 April 2009, 2pp.

Lizarralde, G., Johnson, C., & Davidson, C. (2010) *Rebuilding After Disasters: From Emergency to Sustainability* (Abingdon, UK: Spon Press).

Masci, F., Palangio, P., & Di Persio, M. (2009) Magnetic anomalies possibly linked to local low seismicity, *Natural Hazards and Earth System Sciences*, 9, pp. 1567–1572.

Negri, I. (1996) *La Grande Abbuffata: Mani Rapaci sull'Irpinia del Terremoto* (Milan: ASEFI).

OECD (2010) *OECD Reviews of Risk Management Policies: Italy 2010. Review of the Italian National Civil Protection System* (Paris: Organisation for Economic Co-operation and Development).

Özerdem, A., & Jacoby, T. (2006) *Disaster Management and Civil Society: Earthquake Relief in Japan, Turkey and India* (London: International Library of Postwar Reconstruction and Development).

Pacor, F., Paolucci, R., Iervolino, I., Nicoletti, M., Ameri, G., Bindi, D., et al. (2009) Strong ground motion recorded during the L'Aquila seismic sequence, *Progettazione Sismica*, 3, pp. 55–66.

Palmieri, F. P. (2003) La direzione centrale della difesa civile, *Obiettivo Sicurezza*, 5, pp. 40–43. Available at http://obiettivosicurezza.vigilfuoco.it/archivio.asp.

Papadopoulos, G. A., Charalampakis, M., Fokaefs, A., & Minadakis, G. (2010) Strong foreshock signal preceding the L'Aquila (Italy) earthquake (Mw 6.3) of 6 April 2009, *Natural Hazards and Earth System Sciences*, 10, pp. 19–24.

Papanikolaou, I. D., Lekkas, E. L., Roberts, G. P., McGuire, B., Fountoulis, I. G., Parcharidis, I., & Foumelis, M. (2009) *The 2009 L'Aquila Earthquake: Findings and Implications*. Event Science Report No. 02 (London: Aon Benfield Hazard Research Centre, University College London).

Plastino, W., Povinec, P. P., De Luca, G., Doglioni, C., Nisi, S., Ioannucci, L., Balata, M., Laubenstein, M., Bella, F., & Coccia, E. (2009) Uranium groundwater anomalies and L'Aquila earthquake, 6th April 2009 (Italy), *Journal of Environmental Radioactivity*, 101, pp. 45–50.

Potter, H. (Ed) (2007) *Racing the Storm: Racial Implications and Lessons Learned from Hurricane Katrina* (Lanham, MD: Lexington Books).

RAI News 24 (2009) Terremoti non sono prevedibile, sostiene esperto ONU, *RAI News 24*, 7 April 2009.

Romeo, R., & Pugliese, A. (2000) Seismicity, seismotectonics and seismic hazard of Italy, *Engineering Geology*, 58, pp. 241–266.

Rossetto, T., Peiris, N., Alarcon, J., So, E., Sargeant, S., Sword-Daniels, V., Libberton, C., Verrucci, E., Del Re, D., & Free, M. (2009) *The L'Aquila, Italy, Earthquake of 6 April 2009: A Preliminary Field Report by EEFIT* (London: UK Earthquake Engineering Field Investigation Team).

Spence, W. J., Herrmann, R. B., Johnston, A. C., & Reagor, B. G. (1993) Responses to Iben Browning's prediction of a 1990 New Madrid, Missouri, earthquake, *US Geological Survey Circular*, 1083, pp. 1–248.

Stacul, J. (2007) Understanding neoliberalism: Reflections on the 'end of politics' in northern Italy, *Journal of Modern Italian Studies*, 12, pp. 450–459.

Stallings, M., & Faust, W. (2009) Drafting, revising, and updating local emergency operations plans: The National Response Framework and the Emergency Support Function Annex model, *Journal of Emergency Management*, 7, pp. 11–18.

Tsolis, G. S., & Xenos, T. D. (2010) A qualitative study of the seismo-ionospheric precursors prior to the 6 April 2009 earthquake in L'Aquila, Italy, *Natural Hazards and Earth System Sciences*, 10, pp. 133–137.

Voight, B. (1990) The 1985 Nevado del Ruiz Volcano catastrophe: Anatomy and retrospection, *Journal of Volcanology and Geothermal Research*, 42, pp. 151–188.

Volpini, A. (2009) *L'Ospedale nel Sistema dei Soccorsi Sanitari* (Rome: Dipartimento Nazionale della Protezione Civile).

Yin, R., & Moore, G. (1985) *The Utilization of Research: Lessons Learned from the Natural Hazards Field* (Washington, DC: Cosmos Corporation).

Zollo, A., Marcucci, A., Milana, G., & Capuano, P. (1999) The 1997 Umbria-Marche (central Italy) earthquake sequence: insights on the main-shock ruptures from near-source strong motion records, *Geophysical Research Letters*, 26, pp. 3165–3168.

Using Natural Capital to Reduce Disaster Risk

CAROLYN KOUSKY
Resources for the Future, Washington, DC, USA

ABSTRACT *Natural capital can be used to reduce disaster losses; for example, wetlands store floodwaters and mangroves buffer storm surge. This paper reviews the experience of several US communities that chose to use natural systems to mitigate a variety of risks. The paper discusses the communities' motivations in choosing to use natural capital as opposed to other risk-reduction measures, the policy approaches for doing so, the political and institutional challenges involved, and strategies for overcoming those challenges.*

Introduction

Wetlands have been protected in the middle and upper portions of the Charles River watershed in Massachusetts to mitigate flooding. In Vietnam, mangrove forests are being replanted to protect residents from storm surges. Revegetation is being undertaken on slopes in China to reduce the risk of landslides. New York City has conserved land and aided farmers in reducing environmental damages in the watershed from which it draws its drinking water to minimize health risks.

These are examples of using natural capital to mitigate risks. The term 'natural capital' refers to a stock of natural resources that provides services to people—so-called 'ecosystem services'. Ecosystem services are the benefits people receive from nature (Daily, 1997) and include such things as purified air and water or recreation.[1] Natural capital can be used to either lower the probability of a disaster occurring or reduce damages should an event occur. Examples of the former include stabilizing slopes with vegetation to prevent landslides, using trees to sequester carbon to reduce risks from climate change, and preserving multi-age forest stands to reduce vulnerability to disease and fire. Examples of the latter include protecting wetlands for flood control and mangroves for wave- and storm-surge attenuation. Some of these actions have long been used and incorporated into policy, while others are only recently receiving attention from decision-makers.

This paper presents the results of in-depth case analysis of five communities in the United States that have turned to natural capital to reduce the risks of flooding or of drinking water contamination. The goal of the research was to identify the motivations local governments had for choosing an ecosystem approach, what policy tools they used to do so, as well as the challenges they faced and how these were overcome. While detailed case analysis was only conducted of US communities, where the greater political and economic context in which they operate was similar, findings from an analysis of secondary material on international cases are also included to identify areas of potential similarities and differences in the policy adoption process.

The analysis suggests that many communities turn to natural capital to reduce risks because doing so can be cost-effective and also provide myriad co-benefits. Co-benefits include such things as enhanced recreational opportunities, economic development, or habitat preservation. There are generally three policy approaches for using natural capital to reduce risks: land protection, land-use regulation, and incentive programmes. Using natural capital to reduce risks is not routine, and communities face an array of political and institutional barriers when trying to do so, such as lack of information or opposition from key stakeholders. No barrier need be fatal, however. Communities have used a range of methods to overcome their challenges, such as multi-stakeholder negotiation and the effective use of public entrepreneurs.

The next section of the paper offers a quick note on the methodology and aims of this paper. The following section turns to why local governments may choose a natural-capital approach. Then the three policy approaches of land acquisition, land-use regulation, and incentive programmes are discussed, providing examples of each. The next section discusses four political and institutional barriers and how local governments can to overcome them. The final section concludes.

A Note on Methodology and Objectives

This paper aims to shed light on the intricacies of policy processes (see Seidman, 1998) surrounding the use of natural capital to reduce risks to a community. Such analysis may help raise awareness among practitioners and scholars involved in disaster-risk and ecosystem-services management of the role natural capital can play in mitigating a wide variety of risks. This paper is seeking neither a generalized finding of why some communities use natural capital and some do not nor an exhaustive list of the pros and cons of using natural capital over other risk-reduction approaches. The sample size of analysis is limited, and only communities that have chosen to use this approach are examined. Despite these limitations, the case analysis provides initial insight on the design and implementation of ecosystem-service approaches and suggests several avenues for more in-depth research going forward.

The paper presents the results of detailed case analysis of the use of wetlands to provide flood control in the Charles River watershed in Massachusetts, along the Napa River in California, and along the Truckee River in Reno, Nevada, as well as two examples of watershed protection to reduce the health risks of contaminated drinking water in New York City, New York, and Boston, Massachusetts. As this paper is interested in what motivates new investments in natural systems for the ecosystem services they provide, water purification cases, such as San Francisco and Seattle, where the watersheds were protected generations ago, are not considered. Boston and New York City were chosen for their relatively recent adoption of watershed protection efforts. The wetland cases were chosen to include examples from both coasts.

In these communities, semi-structured interviews[2] (Merriam, 1998) were conducted in 2005 with elected officials, staff in the local government, concerned citizens, and other groups, depending on the circumstances of a particular case. Follow-up research on all the cases was undertaken in the following years. Government documents, newspaper articles, pamphlets, books, and papers were examined in analyzing each case, allowing for triangulation among sources (Eisenhardt, 1989). This literature review of secondary sources was also done on several other cases in the US and abroad to offer greater perspective on the applicability of the findings to other locations and other risks.

Motivation for Using Natural Capital for Risk Reduction

Before a natural-capital approach is chosen, a community must first decide to address a given risk. Often, this occurs after a triggering event raises interest in risk reduction, as discussed in Section 5.2.The case analysis suggests two main motivations for a community to use natural capital over other risk-reduction measures: using natural capital may be more cost-effective, or it could provide additional co-benefits to the community. These reasons could make investment in natural capital more economically justified or more politically attractive than alternatives.

Natural capital has proven to be cost-effective in several instances. For example, in New York City, investments in the watershed to preserve water quality as opposed to the construction of a water-filtration plant, saved billions of dollars (Appleton, 2002; Chichilnisky & Heal, 1998). Along the Charles River, using wetlands to mitigate flooding cost one-tenth the estimated cost of a dam and levee project, and the US Army Corps of Engineers found that the benefits exceeded the costs, even without considering environmental and recreational co-benefits (US Army Corps of Engineers New England Division, 1993; National Research Council Committee on Assessing and Valuing the Services of Aquatic and Related Terrestrial Ecosystems, 2005).

Ecosystem-service provision will not always be more cost-effective than other measures. There may be cases when the cost of using natural systems is prohibitive, particularly if it requires purchasing large amounts of valuable land or difficult and costly restoration work. The political process could also introduce barriers to adopting a natural capital approach, necessitating very high transaction costs in order to overcome them. This, too, could render a built solution preferable. One example comes from Greenville, South Carolina. The two watersheds that supply water to Greenville are completely owned by the Greenville Water System (GWS) and were placed in a conservation easement with The Nature Conservancy in 1993 (National Research Council, 2000). The water did not meet total coliform standards in the early 1990s, prompting the US Environmental Protection Agency (EPA) to demand a filtration system. An analysis found that biofilms in the distribution pipes caused the high levels of total coliform, so the GWS argued the best solution was to flush them. EPA still mandated filtration. In the interest of time and money, that is, high transaction costs, GWS decided not to contest them and began construction on a filtration plant (National Research Council, 2000). While maintaining protection of the watershed could have potentially been more cost-effective than building the filtration plant, it would not be if fighting to do so was time and money intensive.

Opportunity costs may also be prohibitive. For example, in late 2004, Kirksville, Missouri sold more than 400 acres of city-owned land in the watershed supplying its drinking water to dairy farmers. Several council members noted a budget deficit of $700 000 in the city's water fund as the reason for the sale (Webber, 2005). In developing countries, the opportunity costs of preserving natural capital can also be high, especially when residents have very high discount rates. In these cases, the short term gain from destroying the natural capital outweighs the long-term risk reduction benefit of preserving it. In Haiti, for instance, demand for firewood has led to deforestation of slopes and an increase in landslide risk. Hurricane Jeanne caused around 1800 landslide deaths on the island in 2004 (Sudmeier-Rieux *et al.,* 2006). During other times, though, structural approaches for disaster-risk reduction are prohibitively expensive in low-income countries, and using ecosystem services to reduce risks could potentially be an affordable alternative. For instance, Bangladesh is extending mangroves along the shore to reduce storm impacts (Sudmeier-Rieux *et al.,* 2006).

A natural-capital approach may also be undertaken because it provides many co-benefits to a community in addition to reducing a particular risk, such as increased areas for recreation, aesthetic improvements, habitat protection, or the spurring of economic growth. In Reno, Nevada, wetlands along the Truckee River are being restored for flood protection. One key benefit is increased opportunities for outdoor recreation. The recreational benefits, it is hoped, will attract outdoor enthusiasts that could help stabilize tourism revenue and draw young, talented workers to the area (Voyles, 2003). Between 1999 and 2004, Napa invested more than $196 million in the downtown as the flood project increased the aesthetics of the riverfront (Napa County Flood Control and Water Conservation District, 2004). This investment could have ripple effects in the city, drawing in more tourist revenue and creating jobs.

In certain cases, another co-benefit of a natural-capital approach is reducing human and technological errors or failures associated with structural works that are managed by people. Regarding water quality, such failures could include disease outbreaks caused by distribution and filtration systems in poor repair, lack of proper maintenance, inadequately trained operators, or failure of operators to respond to warning signs (Hrudey *et al.,* 2006). Decades ago, when the Boston water system was being established, managers were aware of the possibility of human and technological failures and thought that having clean water to begin with was safer than using polluted water and filtering it, according to S. Estes-Smargiassi (personal interview, 19 September 2005). The same logic can apply to flood protection; while floods could be so large as to overwhelm wetlands (just as levees can be overtopped), wetlands will not fail from human errors or mistakes in design, construction, or maintenance (or lack thereof), as may have happened in New Orleans with its levee system (Kintisch, 2005). On the other hand, some natural systems are very heavily managed and human errors are still a concern. New York City, for instance, has aided farmers in adopting best practices that would reduce water contamination, particularly from manure. In situations such as this, practices could be neglected or adopted incorrectly, again introducing human error.

Considering the co-benefits of a natural capital approach could be enough to make a project economically justified. For example, an analysis of a watershed in South Australia found that when the benefits of other ecosystem services produced by the land-use changes were also considered, the project was cost-effective (Bryan & Kandulu, 2009). Stressing the co-benefits can also sometimes increase political support for a policy. Policy-makers can capitalize on this by framing the policy as one of achieving multiple goals simultaneously, creating new stakeholders. For instance, in Reno, the chairman of

Harrah's Entertainment supports the project (Voyles, 2003) because the restored floodplain improves recreational experiences along the river, attracting tourists and improving livability. J. Ruefer (telephone interview, 1 August 2005), the water resources planning manager in Reno, noted that a successful project requires meeting the needs of all stakeholder groups.

Policy Approaches

Many risks that can be mitigated using natural capital are influenced by the actions of private landowners. That is, private land-use activities can generate a particular type of negative externality: increases in risk levels. As Farber (1996, p. 4) writes, '[A] single decision to drain and till a farm field, fill a wetland or build a parking lot has little measurable impact on flooding. But when combined with thousands of similar decisions over decades the impact can be devastating'. Using natural capital thus often requires governments to work directly with private landowners.[3] There are three main policy approaches to doing this: (1) purchase and management of the land critical to providing the service of risk reduction (either fee simple or purchase of easements), (2) government regulation of land-use activities that might increase risks, and (3) provision of incentives for landowners to adopt certain land-use activities, which could include pricing policies, taxes and subsidies, or payment schemes. All three options have varying strengths and weaknesses.

An example of government purchase of land comes from the Charles River watershed in Massachusetts. Repeated flooding in the area had spurred demand for flood control. In 1965, Congress authorized a feasibility study for a US Army Corps of Engineers flood project. The Corps finished the first part of the study in 1968 and recommended upgrading a dam at the mouth of the river. This dam was completed in 1978. In 1968, the same year the first part of the study was completed, a storm caused severe flooding in the region. The Corps was able to examine the event first hand, finding that in the upper and middle portions of the watershed, the flood crest moved extremely slowly and it ultimately took a month for all the storm water to reach the dam (Chandler & Doyle, 1978). This confirmed their belief that wetlands were effectively controlling flooding in the middle and upper basins, leading the Corps, in the second part of the study, to ultimately recommend fee-simple purchase or conservation easement acquisition of wetlands in the watershed. Congressional authorization for the project came in 1974. After extensive research and modelling to target land purchases, acquisition began in 1977. The project was completed in 1984. It is the only Corps flood project in New England to use wetlands to reduce flood risk.[4]

The benefits of such an approach are clear: the wetlands are preserved in perpetuity to act as reservoirs for floodwaters. Some of the sites along the Charles are now popular locations for picnics, hiking, canoeing, and bird watching. When no other land uses, apart from light recreation, are compatible with providing risk reduction, acquisition may be optimal. Several potential drawbacks, however, could hinder this approach. If it is an area with significant development pressure, land purchase could be prohibitively costly. Land acquisition also depends on landowners' willingness to sell their property. Finally, if providing the risk reductions requires acquisition of very large amounts of land, there could be political outcry. Boston has purchased land in the watershed that supplies its drinking water. Watershed residents became upset that such purchases were limiting their ability to develop and costing them tax revenue. To address these concerns, Boston ratepayers finance payments in lieu of taxes, which are essentially payments to the watershed governments' equivalent to the highest tax rate that could be assessed on the land if it were developed.[5]

As another approach, government could simply adopt regulations to ensure the land-use activities that maximize risk reduction are being undertaken. Boston, Massachusetts, has undertaken such regulation in addition to land acquisition. In 1992, the state passed the Watershed Protection Act, which directed the Metropolitan District Commission, Division of Watershed Management to regulate land use in the watersheds from which the greater Boston area gets its drinking water. The regulations include establishing buffer strips along streams and reservoirs, limiting the amount of impervious surface area, and restricting hazardous material usage.

Regulation can guarantee provision of the desired ecosystem service, but it may require monitoring and enforcement and will not work when the community receiving the benefits of the service is in a different political jurisdiction than the ecosystem generating the service. This was the case for Boston, but the state, which did have jurisdiction, regulated land use in the watersheds. As with large land purchases, another drawback could be political outcry over the regulations.

The third approach is to use incentives to try and induce landowners to engage in behaviour that supplies the desired ecosystem service. Governments could accomplish this through pricing policies, including subsidizing preferable land-use activities or taxing activities that reduce provision of the service. One particular type of incentive programme that is gaining increasing attention is payments for ecosystem services (PES). Wunder (2007, p. 50) has defined PES policies as 'a voluntary, conditional agreement between at least one "seller" and one "buyer" over a well-defined environmental service—or a land use presumed to produce that service'. In practice, the buyer can be a private entity or the government. As most risk-reduction services are public goods that provide benefits to many, it is often government who is the purchaser.

New York City provides a celebrated example. Instead of regulating watershed residents, since this was politically untenable, and in addition to some land acquisition, the city agreed to a voluntary programme[6] in which farmers would be aided in adopting more water-friendly land uses. Farmers in the watershed received improvements to their land financed by the city, which often had the valuable benefit of saving them time from the unpleasant task of handling manure (Appleton, 2002). These benefits to the farmers have been characterized as the 'payments' for the ecosystem service of improved water quality.

There are some cases of private PES programmes; one comes from the bottled water company Vittel. Vittel (now owned by Nestlé Waters) bottles water from an aquifer in northern France. Intensification of farming in the region was threatening the water with contamination from nitrates and pesticides. Relocating to a new area would threaten the brand and label, purchasing all the land was not possible due to French laws restricting sale of agricultural lands for non-agricultural uses, and there was no legal way to compel the farmers to change their behaviour; the company thus turned to PES (Perrot-Maître, 2006). Beginning in 1989, it took 10 years of dialogues, research, and bargaining to establish the current system of contracts between farmers and Vittel. Vittel offers long-term contracts (most are 30 years) which, in return for changes in farming practices that reduce contaminants to the water, offer some abolition of farmer debt, a yearly payment during transition to more sustainable practices, equipment necessary for the new practices, extra labour, and free technical assistance (Perrot-Maître, 2006). This has protected the Vittel water.

As the Vittel case shows, PES and other incentive schemes can be a useful approach when traditional regulatory methods or purchase of the land are not viable options. Incentive programmes are often much more acceptable to all stakeholders since they are voluntary and do not impose costs on non-beneficiaries of the policy. The downside is that when voluntary, exact service levels cannot be predicted, such that provision levels could be too low. For PES in particular, another drawback is that the transaction costs can often be high. As seen in Vittel, often years of negotiation are required to come to an outcome that is mutually beneficial for all involved. To date, most PES programmes have not focused on risks apart from contaminated drinking water, but they could be designed to address a wider variety of risks, such as reducing landslide risk through incentives to grow vegetation on steep slopes (Kiersch, 2008).

Policy Challenges

Policies to provide risk-reducing ecosystem services can face a variety of political and institutional challenges. This section reviews four that emerged repeatedly in the case analysis and literature review: lack of information, uninterested decision-makers, political opposition, and institutional biases and inertia. Examples of how communities have overcome these challenges are discussed.

Lack of Information

The conditions under which ecosystems can reduce risks and the amount of risk reduction gained for various levels and types of conservation activities have been well researched. This research has largely been observational studies of disaster impacts. For instance, the Asian Tsunami of 2004 allowed scientists to observe the protective role mangroves and coastal dunes can play in buffering wave damage (Liu *et al.*, 2005; Danielsen *et al.*, 2005), and Hurricane Andrew allowed for measurements of how coastal wetlands reduce storm surges (Louisiana Coastal Wetlands Conservation and Restoration Task Force and Wetlands Conservation and Restoration Authority 1998). Simulation research has also been conducted (e.g. Wamsley *et al.*, 2009). Such broad scientific information is a necessary first step for policy-makers to be able to adopt a natural-capital approach.

Very specific information on the link between land uses and service levels and about the policy measures to secure such services is also necessary for policy design and adoption. For example, habitat for predators of agricultural pests can be conserved on farms to minimize crop loss, but without careful knowledge of how to do this, practitioners could unintentionally make a pest problem worse (Landis *et al.,* 2000). As another example, several programmes in countries hit by the 2004 tsunami have started programmes to replant mangroves, as they can be localized shields against wave damage. Many of these programmes faced setbacks with seedlings dying because they were planted at the wrong time or in the wrong place or were the wrong species (Check, 2005). In a final example, a local government must not just know that wetlands can be used to reduce floodwaters, but also how to optimize placement of wetlands in a watershed to have the largest effect. Decision-makers' lack of information on the specifics surrounding the use of ecosystem services can hamstring use of natural-capital approaches in areas beyond risk reduction, as well (Turner & Daily, 2008).

Often gaining the necessary technical knowledge requires communities to draw on outside experts. In Napa, for instance, a community coalition created by Friends of the Napa River brought in experts like the prominent hydrologist Luna Leopold to advise the project. M. Johnston-Block (personal interview, 13 June 2005), a member of Friends and the coalition, which included members from a wide range of environmental, government, and community stakeholder groups, noted, 'We built a very strong engineering and scientific group of people who brought us the world's best understanding of rivers and how without destroying the river we could protect the community'. In another example, New York City conducted research and monitoring of water quality in partnership with Cornell University, with the scientific findings supporting changes in farm activities (Smith & Porter, 2009).

It may be the work of a public entrepreneur that overcomes a lack of critical information among decision-makers, as suggested above by the role of Friends in bringing in outside experts. Much theoretical and empirical work has centred on public entrepreneurs, and a variety of definitions have been put forward in previous literature, such as 'people who seek dynamic policy change' (Mintrom, 1997, p. 44), 'innovators who promote new public policy' (Roberts & King 1996, p. 10), and 'advocates who are willing to invest their resources—time, energy, reputation, money—to promote a position in return for anticipated future gain' (Kingdon, 1995, p. 179). Public entrepreneurs are often the ones to bring information on an innovation to a government and can thus be instrumental in overcoming a lack of awareness about a potential solution. This is part of their 'key' role in 'providing high-quality information to decisionmakers' (Mintrom, 1997, p. 44). The public entrepreneur's role as information provider extends to education designed to bolster support for the policy—what Kingdon (1995) terms 'softening up'.

While many risk-reduction activities are undertaken by governments, private entities may also turn to natural capital to reduce risks they face. In this case, overcoming the lack of information could be so costly as to jeopardize the approach. For instance, Integrated Pest Management (IPM) is a technique for preventing infestations in agriculture by using information on the pests and their predators. This quasi-natural capital approach does not rely on conservation but on the functioning of natural systems. It is very information intensive, however, and thus has high up-front costs in terms of learning. As a result, when farmers are uncertain about the benefits of IPM, they have little incentive to test the approach and learn on their own (Cowan & Gunby, 1996). However, the more farmers adopt IPM, the more knowledge is gained that they can spread to others, thus generating a positive externality of adoption. Governments could help internalize these benefits by providing economic incentives to try IPM and public entrepreneurs could help reduce these upfront costs by providing information to individual farmers.

Uninterested Decision-Makers

Politicians have a limited amount of time and attention, and many issues simply do not make it onto their agenda. Preventive risk-reduction measures often fall in this category. As a result, it often takes a triggering event to get an issue on the political agenda. Used here, a triggering event is any external event that prompts interest in adoption of a new policy. Triggering events can capture policy-makers' attention and cause dissatisfaction with the status quo, generating a search for new policies and creating an environment favourable to innovation (Nice, 1994; Sanger & Levin, 1992).

Triggering events for risk reduction are often disasters. For instance, major flood events spurred interest in flood control in Reno, Boston, and Napa. After that occurs, public entrepreneurs can focus attention on natural capital as a solution. Sometimes, though, the nature of the triggering event itself focuses attention on natural capital, as was the case with the 1968 flood in Boston, which allowed the Corps to observe wetlands' protective role (Chandler and Doyle, 1978; US Army Corps of Engineers New England Division, 1976).

Triggering events, however, need not be disasters. Another type is regulation. New York City gets its drinking water from three watersheds that are grouped into two systems—the Cronton River system and the Catskill-Delaware system. For over a century, New York City did not need to filter its water, as these watersheds provided it with pristine water characterized as the champagne of drinking waters (Appleton, 2002). In the 1980s it became clear that the quality of New York City's water was declining. The issuance of the Surface Water Treatment Rule in 1989, which set national water-quality standards for drinking water, was the triggering event for the city to clean up the Catskill-Delaware system using a natural-capital approach.[7] Building a filtration plant was estimated to cost $4–6 billion in up-front capital costs and $250 million annually in operating costs (Appleton, 2002). The city looked to watershed management as a way to save money.

While triggering events can provide an opening to address disaster risk through natural capital, often once a disaster occurs, it may be too late for a natural-capital approach. After a watershed is paved and developed, for example, it is too costly and politically impossible to reclaim and restore the land to a natural state. Luckily, New York City was forced into action before natural capital was ruled out as a viable solution, partly because small declines in water quality are observable, which can trigger investments to prevent further degradation. For other risks, changes in risk levels are unobservable until it is too late. For example, it was not immediately clear that the loss of wetlands in coastal Louisiana was increasing hurricane risk, and restoration is now difficult and extremely costly.

The fact that natural capital provides many co-benefits to communities can be another hook for getting the policy on decision-makers' agendas even as a preventative measure. For instance, local officials may not be overly concerned with protecting wetlands, but if framed as solving a flood problem, promoting economic development, and creating recreational opportunities, as it was in Napa, it is more likely to be attractive to policy-makers. As stated earlier, the co-benefits of a natural capital approach can lead other stakeholders to support a project, increasing the size of the constituency interested in the project. This will not escape the notice of policy-makers. For example, according to Johnston-Block (personal interview, 13 June 2005), Senator Boxer informed Friends of the Napa River that she would be their advocate for the wetland flood project only if the community supported it. The project did have community-wide support due to the myriad benefits it would bring.

Political Opposition

While the co-benefits of using natural capital can create a broad array of supporters, one group, or many groups, may be opposed. A problem arises when a small group of individuals oppose a project that provides overall net benefits. If the group has undue political influence, it could prevent the project from going forward. For example, many poor communities in Indonesia and elsewhere would like to preserve and restore mangroves for the protection they provide from wave and storm damage, but shrimp aquaculture is a powerful interest group that is halting projects (Check, 2005). In this case, the mangroves are providing a local public good of risk reduction. Public goods are non-rival (one person's consumption of the good does not diminish another's) and non-excludable (it is not possible to prevent anyone from consuming the good). Since landowners cannot capture the benefits of protecting the mangroves for mitigating risk, they have no incentive to do so; public goods are under-provided by the market. Public goods are thus often provided by government. Governments face an array of competing values, goals, and stakeholders, however, and the political process may not result in a natural capital approach if powerful groups are opposed.

Reverse cases also occur: the natural-capital approach provides benefits to affluent or politically powerful groups but imposes costs on a small minority. When this minority is vocal, it can force concessions. In New York City, for example, a 1905 state law gave the city the authority to regulate land use in the watershed to protect public health against contaminated drinking water. However, the city's first attempt at regulating the watershed met with enormous resistance from watershed residents. In part, this was bred by years of resentment in the way the city had treated residents when constructing its enormous water supply system (Isakson, 2002). A similar situation occurred in Boston. In both cases, watershed residents became angered when they felt they were being asked to bear the costs of watershed protection (in terms of potentially unprofitable behavioural change or a stifling of the region's economic development), while all the benefits were going to those outside the community. If the policy generates benefits that more than offset the costs, however, mechanisms can be developed to redistribute those benefits.

Such redistribution occurred in both New York City and Boston. In New York City, negotiations between the farmers, the watershed towns, and the EPA ultimately resulted in a Memorandum of Agreement signed in 1997. Farmers in the watershed received improvements to their land financed by the city, and the city, realizing it preferred agriculture to development, is working with the farmers to keep farming in the region viable (Appleton, 2002). Similarly, in Boston, the city provides a variety of benefits to residents of the three watersheds that supply its drinking water. Boston-area ratepayers finance payments in lieu of taxes, allow public access on some of its land for recreation, provide interpretive services to the communities (such as ranger programmes in the schools), and provide technical assistance to the towns for planning and monitoring.

As seen in New York, the possibilities of transfers or side payments to those bearing the costs of a policy can be agreed upon in multi-stakeholder negotiation, and this type of negotiation can also broaden the support base for other natural-capital approaches, such as the use of wetlands for flood control in Napa and Reno. When successful, multi-stakeholder negotiation can create, as it did in Napa, a project on which 'there is not just total, but passionate consensus' (Johnston-Block, personal interview, 13 June 2005). The transaction costs of multi-stakeholder involvement, however, can be high. In Napa, one of the civil engineers noted that 'the total costs of the coalition process and campaign approached $1 million, which does not include the 7000 person hours that the public, resource agencies and City staff contributed' (Wadsworth, 1998, p. 5). In the Vittel case, it took 10 years of mutual dialogue between the farmers and the company to establish trust (Perrot-Maître, 2006).

Several factors can increase the likelihood that negotiations are successful, some of which also lower the costs. Napa used a skilled facilitator to guide the negotiations, improving meetings. Stakeholders also often need to overcome historic tensions (Wondolleck & Yaffee, 2000), as was seen quite clearly in New York City. Having a common goal can help reduce tensions and encourage cooperation (Sherif, 1958). Establishing such a goal often means overcoming what Bazerman (1994) calls a fixed pie bias, or the assumption that interests necessarily and directly conflict. In New York City, stakeholders were able to overcome their bias with the realization that all parties preferred to keep agriculture in the watershed viable. As a result, maintaining rural livelihoods became a twin goal along with preserving water quality (Smith & Porter, 2009). In Napa, the realization that each stakeholder group had a reason for wanting the flood project to succeed became the 'magic glue' that held the coalition together, according to Johnston-Block (personal interview, 13 June 2005). Such partnerships, however, can be fragile. In New York, watershed residents are increasingly concerned about the impact of the city's land acquisition and easement purchases on development in the region (Smith & Porter, 2009).

Biases and Inertia

Many public and private institutions are not accustomed to using natural capital, and institutional biases and inertia may prevent or delay projects. In Napa, voters rejected a US Army Corps of Engineers plan to reduce flooding from the Napa River three times. After a flood in 1995 and the third rejection of the Corps plan, the Friends of Napa River created the community coalition, which helped design a project that removes levees and restores more than 650 acres of tidal wetlands to absorb flood flows. The project was initially very difficult for the Corps to accept. Coalition founding member Johnston-Block (personal interview, 13 June 2005) observed:

> Dave Dickson and MIG [a consulting group] came up with a solution: a tree with two limbs—we called it a two-track approach and said both limbs would be pursued simultaneously. One limb was the old project and the other limb was the new project. . . This double track went along, but from the beginning it was clear that it was just the Corps' security blanket to keep them going while they adjusted to the inevitable. Without it we might have had the war of the watershed. The paradigm change was so huge for the Corps . . . And if we had jammed it down their throats, we might never have had the flood project.

As this observer notes, the use of natural capital was a large paradigm shift, and these can be very difficult for institutions.

Over time, networks form around certain policies and techniques, creating an environment that lowers the costs of finding technical advice, best practices, and support. Such networks can also make moving from the status quo costly. A natural-capital approach to risk reduction will often require a new network of practitioners that includes individuals, such as ecologists, with quite different expertise than are usually involved in disaster management. If a natural-capital approach requires shifting away from a well-supported network, 'excess inertia' could occur, whereby no community or institution is sufficiently motivated to be the first to switch approaches (Farrell & Saloner, 1985). There could be professional inertia, from the challenges of needing new expertise, institutional inertia, from the difficulty in shifting the mission or focus of an institution, or regulatory inertia, where regulations lag behind current understanding. More case research is needed to tease these apart more completely, uncover which are of particular concern regarding the use of natural capital for risk reduction, and determine how each can be most effectively addressed.

Inertia is challenging to overcome, but intermediary groups or individuals—public entrepreneurs—can play a role in creating new networks and providing needed support. In Napa, the coalition took it upon itself to seek out information and advice from informed individuals and groups with whom the Corps did not usually work. As more communities turn to natural capital solutions, biases will begin to break down, and information will be available to help overcome the inertia of changing to a new approach. For instance, Napa is now sharing its substantial expertise in taking a natural-capital approach to flood control with other communities.

Conclusion

The use of natural capital to reduce a variety of risks, from flooding to drinking-water contamination, is one more tool risk managers should have in their toolbox. Scientific information is beginning to build about the possibilities and limits of the use of natural capital to lower the probability or consequences of disasters. To complement this growing scientific understanding, this paper investigates the policy process of using natural capital over more traditional risk management approaches. Lower costs and co-benefits can attract policy-makers to the use of natural capital, and they have a variety of policy mechanisms for achieving risk reductions, but, as with any new approach, they can also encounter difficulties.

This paper highlighted four potential challenges to increasing adoption of the use of natural capital to reduce risks: lack of information, uninterested decision-makers, political opposition, and biases and inertia. Communities have effectively found ways around these challenges by taking advantage of triggering events, working with dedicated public entrepreneurs, and turning to potentially new methods, such as multi-stakeholder negotiation. While these insights should be useful to communities interested in the use of natural capital, these cases' most important offerings may simply be precedent and experience.

Acknowledgements

I would like to thank Lynn Scarlett and two anonymous referees for helpful comments on an earlier draft of this paper.

Notes

1. Definitions of ecosystem services vary. For one definition and classification that is becoming widespread, see Millennium Ecosystem Assessment (2005). For a discussion on a more precise definition for measurement and accounting of such services, see Boyd and Banzhaf (2007).
2. New questions were asked in response to participant answers, in addition to a fixed list of questions.
3. This can be further complicated if the land-use activities that alter risk levels are far away in space or time from the community seeking risk reductions (see Kousky & Zeckhauser, 2006).
4. Personal communication with the Chief of the Planning Branch of the New England District of the US Army Corps of Engineers in 2005 indicated there have been no natural valley storage projects since the Charles River case. There have been, however, other non-structural approaches to flood control. The Corps has also been working since that time in the Northeast with The Nature Conservancy to improve watersheds and more recently is working with them in Reno, Nevada; Sacramento, California; and possibly other places, to integrate flood control with ecosystem restoration.
5. The Department of the Interior makes similar payments to local governments with non-taxable federal lands within their jurisdiction.

6. The newly created Watershed Agricultural Council guaranteed a participation rate of 85%, which was exceeded.
7. The city made the decision to filter the Cronton watershed, which already was being developed, to protect it from non-point source pollution.

References

Appleton, A. F. (2002) *How New York City Used an Ecosystem Services Strategy Carried Out Through an Urban-Rural Partnership to Preserve the Pristine Quality of its Drinking Water and Save Billions of Dollars* (Tokyo: Forest Trends).

Bazerman, M. H. (1994) *Judgment in Managerial Decision Making* (New York: John Wiley & Sons).

Boyd, J., & Banzhaf, S. (2007) What are ecosystem services? The need for standardized environmental accounting units, *Ecological Economics*, 63(2–3), pp. 616–626.

Bryan, B. A., & Kandulu, J. M. (2009) Cost-effective alternatives for mitigating *Cryptosporidium* risk in drinking water and enhancing ecosystem services, *Water Resources Research*, 45, p. W08437.

Chandler, J. P., & Doyle, A. F. (1978) An alliance with nature, *Water Spectrum*, Summer, pp. 24–30.

Check, E. (2005) Natural disasters: Roots of recovery, *Nature*, 438, pp. 910–911.

Chichilnisky, G., & Heal, G. (1998) Economic returns from the biosphere, *Nature*, 391, pp. 629–630.

Cowan, R., & Gunby, P. (1996) Sprayed to death: path dependence, lock-in and pest control strategies, *The Economic Journal*, 106, pp. 521–542.

Daily, G. C. (1997) *Nature's Services: Societal Dependence on Natural Ecosystems* (Washington, DC: Island Press).

Danielsen, F., Sorensen, M. K., Olwig, M. F., Selvam, V., Parish, F., Burgess, N. D., Hiraishi, T., Karunagaran, V. M., Rasmussen, M. S., Hansen, L. B., Quarto, A., & Suryadiputra, N. (2005) The Asian tsunami: A protective role for coastal vegetation, *Science*, 310(5748), pp. 643.

Eisenhardt, K. M. (1989) Building theories from case study research, *The Academy of Management Review*, 14, pp. 532–550.

Farber, S. (1996) *On Borrowed Land: Public Policies for Floodplains* (Cambridge, MA: Lincoln Institute of Land Policy).

Farrell, J., & Saloner, G. (1985) Standardization, compatibility, and innovation, *The RAND Journal of Economics*, 16, pp. 70–83.

Hrudey, S. E., Hrudey, E. J., & Pollard, S. J. T. (2006) Risk management for assuring safe drinking water, *Environment International*, 32(8), pp. 948–957.

Isakson, R. S. (2002) *Payments for Environmental Services in the Catskills: A Socio-Economic Analysis of the Agricultural Strategy in New York City's Watershed Management Plan* (San Salvador, El Salvador: Fundación PRISMA).

Kiersch, B. (2008) Potential of payment for ecosystem services schemes for landslide risk reduction, in: N. Casagli, R. Fanti & V. Tofani (Eds) *Web Proceedings of the First World Landslide Forum*, pp. 706–709 (Tokyo: International Consortium on Landslides and the United Nations International Strategy for Disaster Reductions).

Kingdon, J. (1995) *Agendas, Alternatives, and Public Policies* (New York: Longman).

Kintisch, E. (2005) Hurricane Katrina: levees came up short, researchers tell Congress, *Science*, 310(5750), pp. 953–955.

Kousky, C., & Zeckhauser, R. (2006) JARring actions that fuel the floods, in: R. J. Daniels, D. F. Kettl, & H. Kunreuther (Eds) *On Risk and Disaster: Lessons from Hurricane Katrina*, pp. 59–73 (Philadelphia: University of Pennsylvania Press).

Landis, D. A., Wratten, S. D., & Gurr, G. M. (2000) Habitat management to conserve natural enemies of arthropod pests in agriculture, *Annual Reveiw of Entomology*, 45, pp. 175–201.

Liu, P. L.-F., Lynett, P., Fernando, H., Jaffe, B. E., Fritz, H., Higman, B., Morton, R., Goff, J., & Synolakis, C. (2005) Observations by the international tsunami survey team in Sri Lanka, *Science*, 308(5728), p. 1595.

Louisiana Coastal Wetlands Conservation and Restoration Task Force and Wetlands Conservation and Restoration Authority (1998) *Coast 2050: Toward a Sustainable Coastal Louisiana* (Baton Rouge, LA: Louisiana Department of Natural Resources).

Merriam, S. B. (1998) *Qualitative Research and Case Study Applications in Education* (San Francisco: Jossey-Bass Publishers).

Millennium Ecosystem Assessment (2005) *Millennium Ecosystem Assessment Synthesis Report* (Washington, DC: Island Press).

Mintrom, M. (1997) The state-local nexus in policy innovation diffusion: The case of school choice, *Publius: The Journal of Federalism*, 27(3), pp. 41–60.

Napa County Flood Control and Water Conservation District (2004) *Progress and Plan Summary 2004* (Napa, CA: Napa County Flood Control and Water Conservation District).

National Research Council (2000) *Watershed Management for Potable Water Supply: Assessing the New York City Strategy* (Washington, DC: National Academies Press).

National Research Council Committee on Assessing and Valuing the Services of Aquatic and Related Terrestrial Ecosystems (2005) *Valuing Ecosystem Services: Toward Better Environmental Decision-Making* (Washington, DC: National Academies Press).

Nice, D. C. (1994) *Policy Innovation in State Government* (Ames, IA: Iowa State University Press).

Perrot-Maître, D. (2006) *The Vittel Payments for Ecosystem Services: A 'Perfect' PES Case?* (London: International Institute for Enviroment and Development).

Roberts, N. C., & King, P. J. (1996) *Transforming Public Policy: Dynamics of Policy Entrepreneurship and Innovation* (San Francisco, CA: Jossey-Bass Publishers).

Sanger, M. B., & Levin, M. A. (1992) Using old stuff in new ways: innovation as a case of evolutionary tinkering, *Journal of Policy Analysis and Management*, 11(1), pp. 88–115.

Seidman, I. E. (1998) *Interviewing as Qualitative Research: A Guide for Researchers in Education and the Social Sciences* (New York: Teachers College Press).

Sherif, M. (1958) Superordinate goals in the reduction of intergroup conflict, *The American Journal of Sociology*, 63, pp. 349–356.

Smith, L. E. D., & Porter, K. S. (2009) Management of catchments for the protection of water resources: Drawing on the New York City Watershed experience, *Regional Environmental Change*, 23 October. Available at http://www.springerlink.com/content/l177727548811101/ (accessed 19 April 2010).

Sudmeier-Rieux, K., Masundire, H., Rizvi, A., & Rietbergen, S. (2006) *Ecosystems, Livelihoods and Disasters: An Integrated Approach to Disaster Risk Management* (Gland, Switzerland: The World Conservation Union).

Turner, R. K., & Daily, G. C. (2008) The ecosystem services framework and natural capital conservation, *Environmental and Resource Economics*, 39(1), pp. 25–35.

US Army Corps of Engineers New England Division (1976) *Natural Valley Storage: A Partnership with Nature* (Waltham, MA: US Army Corps of Engineers New England Division).

US Army Corps of Engineers New England Division (1993) *Massachusetts Natural Valley Storage Investigation, Section 22 Study* (Waltham, MA: US Army Corps of Engineers and Commonwealth of Massachusetts, Executive Office of Environmental Affairs).

Voyles, S. (2003) A new era for the Truckee, *Reno Gazette-Journal*, 3 November, p. 3.

Wadsworth, G. S. (1998) Floodplain and watershed management in the Napa Valley, in: Floodplain Management Association (Eds) *Floodplain Management Association Fall Conference: 'After the storm'* (Sacramento, CA: Floodplain Management Association).

Wamsley, T. V., Cialone, M. A., Westerink, J., & Smith, J. M. (2009) *Numerical Modeling System to Simulate Influence of Marsh Restoration and Degradation on Storm Surge and Waves* Coastal and Hydraulics Engineering Technical Note ERDC/CHL CHETN-I-77, Vicksburg, MS: US Army Engineer Research and Development Center.

Webber, M. (2005) Experts: Watershed sale legal, but concerns remain, *Kirksville Daily Express*, 9 January.

Wondolleck, J. M., & Yaffee, S. L. (2000) *Making Collaboration Work: Lessons from Innovation in Natural Resource Managment* (Washington, DC: Island Press).

Wunder, S. (2007) The efficiency of payments for environmental services in tropical conservation, *Conservation Biology*, 21(1), pp. 48–58.

The Dynamics of Informal Responses to Covariate Shocks

R. A. BALGAH & G. BUCHENRIEDER
Leibniz Institute of Agricultural Development in Central and Eastern Europe (IAMO), Halle, Germany

ABSTRACT *Microeconomic theory suggests differential resilience of formal and informal instruments to natural and man-made shocks. Informal risk instruments effective in managing idiosyncratic risks are predicted to collapse in the wake of covariate shocks. This position leads to two hypotheses: (1) that covariate risks are more virulent, thus less resisted by informal instruments, demanding market or state intervention (which are more resilient) to manage aggregate shocks when they occur; (2) that idiosyncratic shocks are less aggressive than covariate shocks, thus can be reasonably accommodated by informal mechanisms. Using empirical evidence, this paper argues that there is no such clear evidence of the superiority of formal instruments over informal responses to covariate shocks. Rather, improper understanding of the dynamic response of informal instruments to aggregate shocks is a prime motive for its underestimation. The paper concludes with suggestions on pertinent research issues that could better capture and explain the role of informal structures in managing shocks.*

Introduction

Microeconomic theory predicts differential resilience of existing instruments to natural and other shocks. Following covariate shocks, informal instruments or safety-net arrangements effective in managing idiosyncratic risks are hypothesized to break down, with formal (public- and market-based) instruments being more resilient to such shocks (Siegel & Alwang, 1999; Holzmann *et al.*, 2003). In spite of increasing evidence of the importance of social capital on economic outcomes (Granovetter, 1973, 1983, 2005; Moody & White, 2003; Goyal, 2005; Miguel & Gertler, 2006; Syrett & Evans, 2007; Bouma *et al.*, 2008; Akçomak & Weel, 2009), the relevance of community safety-nets and risk pooling in managing risks (Sharma & Zeller, 1997; Conning & Kevane, 2002; Van Bastelaer & Leathers, 2006; Park, 2006; Berhanne *et al.*, 2009), microeconomic theory still holds that informal structures are unable to withstand pressures from covariate shocks. Informal arrangements describe the bundle of measures taken at household or community level to protect against risks or to combat shocks, in the absence or presence of public or market based arrangements. Examples include the buying and selling of real assets (such as cattle and gold), informal borrowing, crop diversification, storage, community support arrangement, and kinship arrangements, for example through marriage. Shocks result in income or consumption volatility (Günther & Harttgen, 2009) for which households are not passive. In the case of a drought for example, affected households employ various formal and informal instruments aimed at achieving responsive outcomes such as consumption smoothing (Holzmann *et al.*, 2003), or asset smoothing (Zimmerman & Carter, 2003; Carter *et al.*, 2007).

Formal instruments include both market and public actions against precedented and unprecedented risks or shocks. The term 'risk' is used when referring to uncertain (i.e. stochastic) events and outcomes with known or unknown probabilistic distributions (Siegel & Alwang, 1999; Heitzmann *et al.*, 2001; Alwang *et al.*, 2001; IFPRI, 2002; Sricharoen, 2006). Risk therefore refers to the prospect of a shock (Fafchamps, 2003). Risks are manifested in the form of shocks. Natural shocks (or disasters) emanate from extreme natural events which can be slowly developing (e.g. desertification and epidemics) or could be sudden (e.g. earthquakes, floods, volcanoes). Man-made shocks, on the other hand, comprise events with welfare-depriving capabilities resulting from the breakdown of regular endogenously embedded structures within the social system such as riots, war, and financial crises (Albala-Bertrand, 1993). The focus of this paper is on natural shocks, the impacts of which strongly depend on the household risk management strategies or exposure prior to the shock. In this article (natural) shocks will be used interchangeably with (natural) disasters. Market-based arrangements include financial transactions with formal financial institutions (such as banks and insurance companies), while public arrangements include publicly mandated programmes such as employment and insurance against unemployment, old age, disability, widowhood, and sickness.

Without doubt, however, informal arrangements still constitute the main source of risk management for the majority of the world's population (Holzmann *et al.*, 2003), even if they are conjectured to be somewhat effective in managing idiosyncratic shocks, and disintegrating in the event of covariate shocks. This microeconomic position leads to two major hypotheses.

First, covariate shocks are more virulent than idiosyncratic shocks. Thus households should have a more difficult time coping with covariate as opposed to idiosyncratic shocks. Unfortunately, this is not always the case. Idiosyncratic shocks can create higher welfare losses than covariate shocks. Günther and Harttgen (2009) for example show that in Madagascar, covariate and idiosyncratic shocks have differential effects, with covariate shocks having relatively higher impact on rural households, whereas idiosyncratic shocks impacted more urban households. Estimating the consumption variance at household and community levels as proxy for the impact of idiosyncratic and covariate shocks in Madagascar respectively, they conclude that the 'idiosyncratic variance is much higher than the covariate variance for urban and only slightly higher for rural households. Hence the relative importance of idiosyncratic variance is much higher for urban than rural households' (Günther & Harttgen, 2009, p. 1228). This finding questions the first hypothesis and its underlying microeconomic theory.

Secondly, because formal instruments are more resilient to shocks than informal instruments, they should be applied in the wake of covariate shocks. If for any reason market instruments (such as insurance) are missing, break down or are dysfunctional, then assuming that the state is welfare maximizing, it would intervene through market regulations (in the case of dysfunctional markets) or the provision of public goods and services (in the absence of the market) to buffer against shocks (Mechler, 2004). If the state has a different utility function from its population (as is common in many developing countries) and is not welfare maximizing, then households and communities left only with their 'ineffective', unreliable informal management instruments (Skoufias, 2003) are doomed to uncontainable welfare losses from covariate shocks. Thus theoretically, while formal instruments are assumed effective in managing covariate shocks preventing the shocks to escalate into disasters (with short and long term effects), disaster is eminent in their absence, since informal risk management instruments are ineffective in managing aggregate shocks. But it is well known that in the absence of effective public safety-net systems, poorer households may adopt coping strategies, even if these ultimately prevent them from escaping poverty or reaping the benefits of future economic growth (Skoufias, 2003). This article argues in this direction and beyond.

In the absence of a perfectly functioning state and market, informal instrumental innovations develop that enable agents to become more risk averse, or at least more resilient in buffering shocks. Such processes, dynamic in nature are also active, even in the presence of functioning formal structures. These dynamic processes are known as *adaptation* in the environment literature (see for example Tompkins, 2005), *institutional innovations* or *change* in the new institutional economics literature (Hayami & Ruttan, 1984; Ostrom, 1990; North, 1995), and *systemic process* or *substitution effect* in the natural disaster literature (Albala-Bertrand, 1993, Skoufias, 2003). The large emerging literature on resilience and adaptation to natural disasters (e.g. Holling, 1996; Masten, 2001; Klein *et al.*, 2003; Rose, 2004, 2007; Adger *et al.*, 2005; Berkes, 2007; Gaillard, 2007; Cutler *et al.*, 2008; Carpenter *et al.*, 2009; Heltberg *et al.*, 2009) has not sufficiently addressed the differential roles of formal and informal instruments to buffer idiosyncratic and covariate shocks. In addition, only a blurred boundary exists between idiosyncratic and covariate shocks, as the latter may have characteristics of the former and vice versa. For example, it is difficult to say whether the death of a household member is an idiosyncratic or covariate shock, as death might have occurred because of age—an idiosyncratic shock—or because of an epidemic—a covariate shock (Günther & Harttgen, 2009). In addition, there is no clear cut boundary at which HIV/AIDS for example should be labelled either as an idiosyncratic or covariate shock.

This article argues that the dynamism of informal arrangements has not been sufficiently understood and explored within the economics of natural disasters, although there are a few exceptions (e.g. Albala-Bertrand, 1993, 2000; Skoufias, 2003; Tompkins & Adger, 2004). Norms of reciprocity for example could be transformed into 'wilful sharing' of relief aid, a form of solidarity vital for urgent aid to mitigate the negative impacts of natural disasters. The effective management of shocks with informal and formal instruments demands a better understanding of the former and its dynamic processes, for an optimal combination with the latter. Ex-ante analysis can be more rewarding, although ex-post dynamic examinations can also be helpful (Krackhardt & Stern, 1988; Conway, 2001). The social risk management framework is one means to comprehend and incorporate the dynamism of informal instruments in ex-ante and ex-post management of shocks.

The rest of this paper continues as follows. Section two briefly reviews responses to covariate shocks. Section three discusses the social risk management framework, portraying its usefulness in capturing the dynamism of informal and formal shock management instruments for the proper management of large shocks. Section four presents some empirical case studies that demonstrate the dynamism, resilience, and effectiveness of informal instruments against aggregate shocks. These examples show that informal arrangements do not simply collapse in the wake of covariate shocks as predicted by theory, but are dynamic enough to take different forms in the presence or absence of formal mechanisms. Section five concludes the article with some suggestions on future research directions to further enlighten policy in the management of large natural shocks.

Covariate Shock Responses and Microeconomic Theory: A Brief Review

In order to obtain a better perspective on the dynamic informal instrumental responses to aggregate shocks, it is worthwhile to briefly overview related concepts and terminology as presented in the microeconomic literature.

Shock or disaster response defines a wide array of endogenous and exogenous reactions, measures, and policies that are aimed at mitigating, counteracting, and preventing shock impacts and effects. Impact effects can also stimulate anticipatory or innovative responses aimed at prevention and mitigation of future potential shocks (Albala-Bertrand, 1993, 2000). According to Albala-Bertrand (2000), shock responses generate three main areas of attention: (1) response mechanisms; (2) compensatory response and (3) anticipatory response. Response mechanisms constitute exogenous and endogenous shock response processes. Endogenous response mechanisms are channelled through society's built-in institutional processes: formal and informal feedback mechanisms which are part of the existing self-regulatory social machinery. For example, extended family solidarity represents an endogenous reaction to shocks. Exogenous mechanisms constitute *ad hoc*, unpatterned, unguaranteed, or irregular process expressed through actions, measures, and policies that formally fill gaps left by built-in responses; bypass endogenous channels; shift initiatives away from regular actors, or superimpose alternative structures (Albala-Bertrand, 2000). Thus exogenous mechanisms constitute private and public interventions that go beyond built-in community-based endogenous mechanisms. In the midst of large shocks, various combinations of endogenous and exogenous response mechanisms may emerge, aimed at restituting, reconstructing, anticipating or preventing the effects of shocks. Also, a transformation of existing response mechanisms into situation-appropriate arrangements may develop through dynamic processes. As an example, solidarity and reciprocity seem to have contributed to self organization for members of the Lampoon community in Thailand, as observed a few months after the 2004 tsunamis. In combination with endogenous financial resources, the community was able to optimize the large wave of external (exogenous) aid as a result of the tsunamis to construct permanent residences for all 30 surviving households as part of long-term recovery strategy (Santhanboa, 2008). This example depicts the combination of endogenous and exogenous mechanisms for reconstruction.

Compensatory response describes the reaction to actual shock impact effects to counteract the effects on people, and the economy (emergency) or to restore permanently the losses from direct shock effects on people, stocks and institutions (restitution). Thus while emergency response activities seek to put people back on their feet via both relief and rehabilitation activities, reconstruction aims not at replicating actually damaged items, but at the restitution of the system's lost capabilities, functions and institutions (Albala-Bertrand, 1993, 2000). System elements are not passive in this process. Rather an interplay of dynamic processes of change emerge that determines a new outcome which might or might not exactly mirror the initial status quo, but that at least supports agents to compensate for the negative effects of shocks.

Anticipatory response represents the series of actions and measures that seek to prevent, soften or mitigate future disaster and/or its effects. This response aims at tackling pre-disaster factors (such as societal vulnerability, brittleness, and frailty), the actual impact or its potential effects if the impact cannot be prevented (Albala-Bertrand, 1993, 2000; Smith, 1992).

The brief discussion on shock responses above reflects the ideas embedded in the World Bank's Social Risk Management (SRM) framework. While response mechanisms reflect formal and informal arrangements applied in managing anticipated and unanticipated risks, compensatory responses relate to coping strategies and anticipatory responses describe preventive, reductive and mitigating strategies as used in the SRM literature. These aspects constitute the main elements of the social risk management framework (Holzmann & Jorgensen, 1999, 2000; Holzmann *et al.*, 2003). Thus if well implemented, the SRM framework could provide information not only on the response mechanisms and forms of responses but also on the nature of the shock and the dynamism of the response of formal and informal processes to different types of shocks.

Social Risk Management: Capturing Dynamic Responses to Covariate Shocks

The SRM framework was developed by the World Bank over the past 10–15 years to facilitate risk and vulnerability assessments especially in developing countries where the exposure of households to idiosyncratic and covariate shocks often results in great and sometimes irreversible welfare losses (Siegel & Alwang, 1999; Holzmann & Jorgensen, 1999, 2000; Holzmann *et al.*, 2003). The advent of SRM enhanced a paradigm shift in social protection from a payout phenomenon towards a 'Teaching the man to catch fish' doctrine (Swidler & Watkins, 2009, p. 1), aimed at stimulating sustainability in the social protection domain. Prior to the SRM framework, social protection was broadly defined as public measures to provide income security to individuals and the entire population (Holzmann & Jorgensen, 2000; Holzmann *et al.*, 2003). With the SRM lens, social protection became better understood as 'public interventions to (1) assist individuals, households and communities better manage risks, and (2) provide support to the critically poor' (Holzmann & Jorgensen, 2000, p. 3). This form of assistance demands a better understanding and incorporation of local endowments or capabilities in the process of social protection (see for example Sen (1983, 1985) for further discussion). The World Bank's SRM framework has been successfully applied empirically for risk and vulnerability assessments in many developed and developing countries. Examples include China (Jalan & Ravaillion, 1999), Indonesia (Chaudhuri *et al.*, 2002), Bangladesh, Ethiopia, Mali, Mexico, and Russia (Skoufias & Quisumbing, 2004), and more recently, Madagascar (Günther & Harttgen, 2009). Social risk management can therefore be seen as a strategic move towards sustainable social protection. This innovation was triggered by a number of pertinent issues. First, in OECD-type economies, where social protection programmes such as active market policy, social insurance, and social assistance do exist, the high and often rising public expenditure levels generate concern, particularly in view of an aging population, rising international competition, and increasing shocks. In contrast, developing countries have few public resources and can spend little for the income security of their populations, despite the high levels of poverty and income insecurity of individuals in the formal and informal labour markets (Holzmann & Jorgensen, 2000). A new form of social protection was therefore necessary that took into consideration endogenous risk management capacities and includes aspects of these in social protection policies as prerequisite to enhancing long term sustainability. SRM describes the capacity of a system, community, or society potentially exposed to hazards to adapt, by resisting or changing in order to reach and maintain an acceptable level of functioning and structure. This is determined by the degree to which the social system is capable of organizing itself to increase its capacity for learning from past experiences for better future protection and improvement of risk reduction measures (ISDR, 2008). To be effective the SRM framework should combat information asymmetry, that hitherto led to moral hazard, high transaction costs, non-exogenous risks, inappropriate design and selection of management instruments, use of information as a commodity and instrument to power, and causing market and government failures in the provision of risk management instruments (Holzmann & Jorgensen, 2000; Okuyama, 2003). In other words, social risk management should enhance the functioning of both formal and informal risk management instruments. Therefore, by incorporating measures of a socio-ecological system's ability to recover from shocks and stresses (that is informal instruments); social risk management provides information on community resilience. At the same time, it determines the efficiency and strength of adaptive and coping strategies within the system and enhances the functioning of formal risk management instruments (Singh, 1996). This 'avoid[s] creating a culture of dependency among beneficiaries' (Skoufias, 2003, p. 1093) in long term social protection programmes. Disaster risk reduction represents a new paradigm from traditional disaster response, and involves significantly more than the traditional perspective of improving operational capacities to respond to an impending disaster or providing relief assistance. Essentially, it connects disaster recovery with resilience, with increasing attention on the capacity of affected communities to 'bounce back' with little or no external assistance (Manyena, 2006). Risk reduction reduces vulnerability to natural hazards (Thomalla *et al.*, 2006). Thus, while the need to ensure timely and effective response to an emergency is essential, in many regions, political commitment and resource allocation to address hazardous conditions has been concentrated on short-term emergency contingencies, to the detriment of long-term hazards planning. However, building disaster resilient communities will require more systematic and integrative efforts to protect ecosystems and livelihoods (environmental, social, and material assets) before the actual onset of disaster events (Chang & Shinozuka, 2004).

The *main elements* of the SRM framework consist of risk management strategies; risk management arrangements and level of formality; and actors in risk management. Risk management strategies can be preventive, mitigating, or coping. Preventive strategies are implemented before a risk event occurs to evade, or reduce the probability of a downside risk or welfare loss. Examples include vaccination and resettlement. Mitigation strategies decrease the potential impact of a future downside risk. Like preventive measures, they are implemented before the occurrence of a risk. However, they are not aimed at stopping the risk from occurring, rather to reduce the negative impacts, once the risk occurs or to consolidate positive effects of risks. Mitigation strategies can take the form of portfolio diversification, by relying on a variety of assets from which returns are not correlated to reduce variability in income; informal and formal insurance mechanisms (such as risk pooling and insurance premiums) and hedging (e.g. proper marriage arrangements, precautionary savings, and social networking). Preventive and mitigating strategies constitute ex-ante measures. Coping strategies are measures designed to relieve the impact of risks once they occur. Coping is achieved through self insurance (e.g. selling, pledging, or mortgaging of household assets); informal insurance (such as borrowing from relatives, friends, and money lenders); using market-based support systems (such as bank loans) and external support (such as relief aid) (Holzmann & Jorgensen, 1999, 2000; Holzmann *et al.*, 2003; Mechler, 2004; Berhanne *et al.*, 2009).

Risk management arrangements can be informal, market-based, or publicly mandated (Skoufias, 2003) or better still, a proper mix of different arrangements as perceived, defined, and captured by the SRM framework. As such informal instruments will only be relegated as ineffective if revealed through analysis. The social risk management framework provides a detailed understanding of formal and informal mechanisms as a prerequisite towards better risk or shock management, and enhanced social protection. As demonstrated in the next section empirically, the disintegration of informal instruments against covariate shocks is not automatic as stipulated by theory. Rather, informal instruments can be resilient to both idiosyncratic and covariate shocks.

Informal Instrumental Resilience to Aggregate Shocks: Empirical Evidence

This section deals with a collection of case studies from different continents that illustrate the robustness or resilience of informal instruments to covariate shocks. An example from the Tsunami in Southeast Asia will be presented followed by case studies of informal institutions in Armenia, social networks in Uzbekistan and Kenya, and informal responses to hurricanes Mitch in El Salvador and Katrina in New Orleans.

Reconstruction After the 2004 Tsunami in Bahn Lampoon, Northern Thailand

The Tsunami of December 2004 are the most devastating covariate shocks that have impacted some Asian countries, after the financial crises of the 1990s. In northern Thailand for example, it rendered most inhabitants poor, creating losses that affected the whole Thai economy. It affected six provinces along the Andaman Coast, leaving 8,000 people dead. It impacted 407 villages, completely destroying 47 of them including well-known tourist destinations. Vulnerable fishing communities were destroyed and livelihoods lost with survivors left to cope with psychological trauma. Coral reefs and coastal habitats were destroyed and the intrusion of sea water affected agricultural productivity. The estimated financial loss was US$2 billion, and the overall GDP growth of Thailand reduced by 0.4% (Segschneider & Worakul, 2007).

The Lampoon community in Takua Pa District was seriously affected by the Tsunami. Only 30 of 52 original families could be identified in the community after the disaster. Although a huge wave of aid swept into the affected communities after the retreat of the Tsunami, 'many people in Lampoon had little opportunity to obtain any of this aid. The land [had been] cordoned off and many people were unable to enter and find the remains of their dead ones' (Santhanboa, 2008, p. 9). The Lampoon community is not a fishing community, although it is located along the Andaman Coast. Its inhabitants are primarily former mining workers who migrated into the region in the 1970s but remained after the mining concessions had expired. In spite of land reforms in Thailand in the later part of the 20th century, these villagers could not acquire land titles because the major mining companies rebuked such attempts claiming ownership over the land. In addition, land title issuing officers required bribes which community members were too poor to afford. Thus surviving households in Lampoon community were not only faced with the difficulty to access relief aid and identify their plots, but also with the problem of returning to this property after the disaster even if individual parcels could be identified. Tourism, the major source of income, had been destroyed by the natural shock and they were threatened and prohibited by the mining companies from rebuilding their houses. Only community mobilization and collective action brought progress.

After a series of community meetings, four regional slum networks were formed. Villagers returned to the area on 4 February 2005 with the intention of reconstructing their houses. Fast, quality, and cooperative work was needed, in addition to funding. Through community dialogue, each household agreed and contributed a lump sum of 1,000 Baht for daily expenditures and a collective loan of 500,000 Baht for high cost activities and management (Santhanboa, 2008). US$1 at the time was equivalent to 34.1 Thailand Baht (THB). The community decided on the construction of 30 houses (i.e. one house per surviving family). Some additional financial resources were obtained from non-governmental organizations such as The Thai Red Cross Society, Oxfam, World Vision, and Plan International. Based on a budget drawn up and transparently managed by the stakeholders themselves, and with the employment of community labour, 30 houses were constructed in six months. As one of the villagers recalls:

> During those days, none of us had a good night's sleep. After we started, more and more people began to join in. The rebuilding of houses was intended to become a community activity in order to strengthen the process [of reconstruction].
>
> (Pi Yupin as quoted by Santhanboa, 2008, p. 13)

This example illustrates the role of informal mechanisms in coping with shocks. Whether the Lampoon community inhabitants have completely recovered from the shock is a different question. They organized themselves to initiate recovery processes from a covariate shock, in a situation where state and market instruments were ineffective in providing timely relief aid. Thus, if state and market mechanisms fail, leaving households only with informal instruments, they could be transformed into vital instruments to manage covariate shocks. This example clearly demonstrates how informal safety nets and cognitive social capital (such as reciprocity) can be resilient to aggregate shocks.

Using Informal Institutions for Capital Accumulation in Armenia

In the 1990s, chronic public financing problems pushed health and education to become *de-facto* fee-for-service programmes in Armenia. Parents and children in public schools had to pay for food, instruction, and textbook use fees, as well as for medical services. If nothing was done, the financial shock would further expose the Armenian population to the risk of future aggregate shocks such as epidemics and shortage of quality human labour. Some measures to prevent future risks turning into shocks due to the national financial shock were necessary. The government established a school textbook waiver programme that allocated to each school a fixed amount sufficient to wave only 10%, leaving 90% to be raised by parents by charging a rental fee of about US$1 for each textbook the child uses. Armenia's continuing success in developing human capital through this system has only been possible with the support of informal institutions like parent–teacher associations. These informal institutions contributed to buffering the national financial shock from escalating in

Armenia, in a way that could not be achieved solely through state or market institutions (Conning & Kevane, 2002). Contrary to the findings of Jacoby and Skoufias (1997) in rural India, or Skoufias and Parker (2002) in urban Mexico where large shocks affected school attendance, similar observations were not reported in Armenia. Rather, the Armenian case study supports the notion of Trosper (2002) that highly resilient communities demonstrate essentially their ability to organize, mobilize, and to act collectively.

Informal Networks and Water Distribution in Uzbekistan

Water shortages in drought years can lead to widespread hunger in affected communities. For farmers in Khorezm province in Uzbekistan, this will immediately result in substantial disruption of livelihoods, considering that they are closer to the desert and therefore prone to water scarcity. Following independence in 1991, the centrally planned water management systems collapsed as 'over 75% of managers lost their positions within three years' (Wegerich, 2004, p. 345). State and market institutions were failing, leaving farmers to rely on their own informal networks. Wegerich reveals innovative network structures developed by farmers for communication and water distribution in drought years (for a full discussion on the network structure, see Wegerich, 2004). The results are similar to the earlier findings of Campbell (1999) that provides clear evidence of over 20 years of dynamic responses to drought by farmers and herders prone to recurrent drought-related food security in southern Kenya. These examples illustrate that innovations in informal instruments can induce successful management of covariate shocks, when formal structures fail or are not functional.

Children and Youth Disaster Risk Communication Networks: El Salvador and New Orleans

A disproportionate number of women and children are killed by disasters of a covariate nature (Cutter, 1995; Fordham, 1999). However, ex-ante or ex-post measures continue to assume these groups as passive victims, with little or nothing to contribute in communicating risks or preventing or responding to shocks (Mitchell *et al.*, 2008). Children and youth are not integrated into the risk market and thus should be protected by the state since they lack management instruments, or because their informal instruments cannot withstand pressures from aggregate shocks.

As the case of El Salvador will show, such perceptions can influence informal instrumental performance. As evident in the case study, changes in institutions and perceptions towards the role of children and youths and their non-existent or ineffective informal instruments can be triggered by large shocks.

The inundations caused by Hurricane Mitch in 1998 and the earthquakes of 2001 followed by Hurricane Stan in 2005 affected a total of over 2 million people in El Salvador. Children of poor families were particularly vulnerable (Mitchell *et al.*, 2008). Plan International NGO initiated a programme that raised children and youth awareness to natural disasters in El Salvador. Children and youth were able to identify and participate in the prevention of covariate shocks. As an example, a children's group in El Carrizal municipality identified unregulated quarrying of stone and sand from rivers as major risk, leading to increased erosion and vulnerability to flooding: a covariate risk. As reported by Mitchell *et al.* (2008, p. 261):

> Together, and initially without adult support and despite many adults objecting, they devised a campaign lobbying with their parents and local government authorities. They blockaded roads to the river, pleaded with lorry drivers, erected signs warning of danger, pressured their parents to stand up against quarrying and persuaded local authority to enforce regulations. Quarrying along vulnerable stretches of the river bank has now stopped.

Although the children were exposed to capacity building events from Plan International, it is essentially their ability to organize, mobile, and to act collectively that contributed to the meaningful steps towards reducing the possibility of further shocks. Thus learning-based processes brought about resilient informal institutions that reduced the occurrence of shocks, in a manner that is antithetical to existing macroeconomic theory.

Hurricane Katrina of August 2005 killed over 1000 persons and flooded approximately 80% of the city of New Orleans within 18 hours (US White House, 2006). The following communities were highly affected: African-American families (living close to the poverty line), undocumented Latino labourers, and a Vietnamese community. The Federal Emergency Management Agency (FEMA) did not execute its task with diligence and there was severe lack of trust among residents towards government officials and formal disaster recovery and emergency agencies (Mitchell *et al.*, 2008). The Vietnamese community was observed to exhibit exceptional resilience and the youth and children demonstrated stronger agency. This resilience has been attributed to their 'unique history that stems from the Vietnam War' (Mitchell *et al.*, 2008, p. 267). As Mitchell *et al.* explain further:

> During the hurricane, many adolescent and children assisted in evacuation, relief and recovery. In the recovery period they have been instrumental in boosting [community] morale and bringing to attention other risk issues affecting their community. For example youth members have been participating in the political arena to ensure the community is aware and involved in formal recovery mechanisms. (2008, p. 267)

The recovery process after the shock from Hurricane Katrina in New Orleans, especially in the Vietnamese community, has been enhanced by informal youth groups and community members. This case study re-emphasizes the resilience and dynamism of informal risk management instruments when formal instruments are ineffective. Consolidation of local spaces and expansion of spaces of engagement was achieved through a solidification and appropriation of past experiences.

Conclusion

This aim of this paper is to stimulate discussion on the importance of informal structures especially with respect to covariate natural shocks. Using empirical examples, this paper challenges the existing microeconomic theoretical supposition of robust formal instruments and ineffective informal mechanisms in the wake of large shocks. This depends to a great extent on the effectiveness of specific household and community shock response mechanisms. The failure of the state and market as motivation for informal institutional innovations is recurrent in our case studies (particularly evident in the example from Uzbekistan). Furthermore, in line with existing economic theories of the state and markets, the case studies generally indicate that other dynamic, adaptive, or systemic processes operate within informal mechanisms when exposed to natural shocks. These issues relate to trends well-documented and deeply analyzed in the new institutional economics literature. These trends need to be extended in testing present theories, developing new theories, and implementing policies on microeconomics of shocks. Meanwhile, the SRM framework provides an opportunity for explaining the puzzles surrounding instrumental responses to correlated shocks, with its capability improving when applied ex-ante. Although the framework has been successfully applied for risk and vulnerability assessments in developing and developed countries, it needs to be adapted to capture the strengths, weaknesses and particularly the dynamic processes accompanying both formal and informal responses to aggregate shocks. The search for a better theory demands contemporary research to empirically focus on understanding under what circumstances informal instruments are resilient to covariate shocks, or which factors best explain why informal mechanisms are effective against idiosyncratic risks and not often against covariate risk as predicted by theory. The frequency, correlation, and intensity of shocks have been identified in the literature as determinants of informal responses to shocks (e.g. Holzmann & Jorgensen, 2000; Holzmann *et al.*, 2003; Günther & Harttgen, 2009). The case studies presented in this paper suggest that past experiences or prior exposure to covariate shocks (e.g. the Vietnamese community in New Orleans and farmers in southern Kenya); capacity building and ability to organize (e.g. youths and children in El Salvador); proper recognition of the role of informal institutions (e.g. parent teacher associations in Armenia); and the nature of the shock and lobby or force are additional determinants of robustness of informal instruments against covariate shocks. Nevertheless a closer analysis of the dynamic processes operating in formal and informal settings would bring new insights and a better understanding of the microeconomics of instrumental response to natural and man-made shocks. As remarked by Skoufias (2003), both household and community characteristics matter in differential responses to aggregate shocks. The SRM framework can significantly contribute to this new development as long as it is adjusted to capture and better understand the underlying factors. A better understanding of adaptive informal shock management strategies engrained in iterative learning processes will improve the employment of informal instruments as standalone measures, or complementary to formal instruments, for improved resilience especially to sudden large shocks with welfare reducing capabilities. This hopefully should bring more clarity on the role of informal instruments and initiate new forms of collaborative management, re-orientate the theoretical discourse and policy choices within the microeconomics of natural shocks. The expected outcomes of these processes include robust disaster management plans capable of coping with rapidly changing circumstances; incorporating disaster risk reduction and social disaster resilience into broader issues of ecosystems integrity and sustainable livelihoods, and promoting disaster awareness through appropriate education.

References

Adger, W. N., Hughes, T. P., Folke, C., Carpenter, S. R., & Rockström, J. (2005) Social-ecological resilience to coastal disasters, *Science*, 309(3005), pp. 1036–1039.

Akçomak, I. S., & Weel, B. (2009) Social capital, innovation and growth: Evidence from Europe, *European Economic Review*, 35(5), pp. 554–567.

Albala-Bertrand, J. M. (1993) *The Political Economy of Large Natural Disasters: With Specific Reference to Developing Countries* (Oxford: Clarendon Press).

Albala-Bertrand, J. M. (2000) Responses of humanitarian emergencies and natural disasters: An analytical comparison, *Third World Quarterly*, 21(2), pp. 215–227.

Alwang, J., Siegel, P. B., & Jorgensen, S. L. (2001) *Vulnerability: A view from different disciplines*, Social Protection Discussion Paper No. 0115, Washington, DC: The World Bank.

Berhanne, G., Gardebroek, C., & Moll, H. A. J. (2009) Risk-matching behavior in microcredit group formation: Evidence from northern Ethiopia, *Agricultural Economics*, 40(2009), pp. 409–419.

Berkes, F. (2007) Understanding uncertainty and reducing vulnerability: Lessons from resilience thinking, *Natural Hazards*, 41, pp. 283–295.

Bouma, J., Bulte, E., & van Soest, D. (2008) Trust and cooperation: Social capital and community resource management, *Journal of Environmental Economics and Management*, 56(2), pp. 155–166.

Campbell, D. J. (1999) Response to drought among farmers and herders in Southern Kajiado District, Kenya: A comparison of 1972–1976 and 1994–1995, *Human Ecology*, 27(3), pp. 377–416.

Carpenter, S. R., Folke, C., Scheffer, M., & Westley, F. R. (2009) Resilience: Accounting for the noncomputable, *Ecology and Society,* 14(1), p. 13. Available at http://www.ecologyandsociety.org/vol14/iss1/art13.

Carter, M. R., Little, P. D., Mogues, T., & Negatu, W. (2007) Poverty traps and natural disasters in Ethiopia and Honduras, *World Development*, 35(5), pp. 835–856.

Chang, S. E. & Shinozuka, M. (2004) Measuring improvements in the disaster resilience of communities, *Earthquake Spectra*, 20(3), pp. 739–755.

Chaudhuri, S., Jalan, J., & Suryahadi, A. (2002) *Assessing Household Vulnerability to Poverty from Cross Sectional data: A Methodology and Estimates from Indonesia*, Discussion Paper No. 0102-52, New York: Columbia University.

Conning, J., & Kevane, M. (2002) Community-based targeting mechanisms for social safety nets: A critical review, *World Development*, 30(3), pp. 375–394.

Conway, S. (2001) Employing social networking to reveal tensions between informal and formal organizations, in: O. Jones, S. Conway, & F. Steward (Eds) *Social Interaction and Organizational Change*, pp. 81–124 (London: Imperial College Press).

Cutler, S. L., Barnes, L., Berry, M., Burton, C., Evans, E., Tate, E., & Webb, J. (2008) A place-based model for understanding community resilience to natural disasters, *Global Environmental Change*, 18(4), pp. 598–606.

Cutter, S. L. (1995) Race, class and environmental justice, *Progress in Human Geography*, 19(1), pp. 111–122.

Fafchamps, M., & Lund, S. (2003) Risk sharing networks in rural Philippines, *Journal of Development Economics*, 71(2), pp. 261–87.

Fordham, M. (1999) Participatory planning for flood mitigation: models and approaches, *Australian Journal of Emergency Management*, 13(4), pp. 27–34.

Gaillard, J.-C. (2007) Resilience of traditional societies in facing natural hazards, *Disaster Prevention and Management*, 16(4), pp. 522–544.

Goyal, S. (2005) Strong and weak links, *Journal of the European Economic Association*, 3(2–3), pp. 608–616.

Granovetter, M. S. (1973) The strength of weak ties, *American Journal of Sociology*, 78(6), pp. 1360–1380.

Granovetter, M. (1983) The strength of weak ties: A network theory revisited, *Sociological Theory*, 1(1983), pp. 201–233.

Granovetter, M. (2005) The impact of social structure on economic outcomes, *Journal of Economic Perspectives*, 19(1), pp. 33–50.

Günther, I., & Harttgen, K. (2009) Estimating household vulnerability to idiosyncratic and covariate shocks. A novel method applied in Madagascar, *World Development*, 37(3), pp. 1222–1234.

Hayami, Y., & Ruttan, V.W. (1984) Towards a theory of induced institutional innovations, *Journal of Development Studies*, 20(4), pp. 203–223.

Heitzmann, K., Canagarajah, R. S., & Siegel, P. B. (2001) *The Sources of Vulnerability: A Rationale and Guideline for the Assessment of Risks and Risk Responses*, Social Protection Discussion Paper series, Human Development Network, Washington, DC: The World Bank.

Heltberg, R., Siegel, P. B. & Jorgensen, S. L. (2009) Addressing human vulnerability to climate change: Toward a 'no regrets' approach, *Global Environmental Change*, 19(2009), pp. 89–99.

Holling, C. S. (1996) Engineering resilience verses ecological resilience, in: v-P. C. Schulze (Ed) *Engineering with Ecological Constraints*, pp. 31–44 (Washington, DC: National Academy Press).

Holzmann, R., & Jorgensen, S. (1999) *Social Protection as Social Risk Management. Conceptual Underpinnings for the Social Protection Sector Strategy Paper* (Washington, DC: The World Bank).

Holzmann, R., & Jorgensen, S. (2000) *Social Risk Management: A New Conceptual Framework for Social Protection and Beyond*, Social Protection Discussion Paper No. 0006, Washington DC: The World Bank.

Holzmann, R., Sherburne-Benz, L., & Telsuic, E. (2003) *Social Risk Management. The World Bank's Approach to Social Protection in a Globalized World* (Washington, DC: Social Protection Department, The World Bank).

IFPRI (2002) Background information for presenters in improving understanding of risks and household responses to risk. Paper presented at the World Bank Conference on 'Risk and Vulnerability: Estimation and Policy Implications', 23–24 September, Washington, DC, IFPRI.

International Strategy for Disaster Reduction (ISDR) (2008) *Terminology: Basic Terms of Disaster Risk Reduction* ISDR Secretariat. Available at www.unisdr.org/eng/library/lib-terminology-eng-p.htm (accessed 5 August 2009).

Jacoby, H., & Skoufias, E. (1997) Risk, financial markets and human capital in a developing country, *Review of Economic Studies*, 64(3), pp. 311–335.

Jalan, J., & Ravaillion, M. (1999) Are the poor less well insured? Evidence on vulnerability to income risk in rural China, *Journal of Development Economics*, 58(1), pp. 61–81.

Klein, R. T. J., Nicholls, R. J., & Thomalla, F. (2003) Resilience to natural hazards: How useful is this concept? *Environmental Hazards*, 5(2003), pp. 35–45.

Krackhardt, D., & Stern, R. N. (1988) Informal networks and organizational crises: An experimental simulation, *Social Psychology Quarterly*, 5(2), pp. 123–140.

Masten, A. S. (2001) Ordinary magic: Resilience processes in development, *American Psychologist*, 56(3), pp. 227–238.

Manyena, S. B. (2006) The concept of resilience revisited, *Disasters*, 30(4), pp. 434–450.

Mechler, R. (2004) *Natural Disaster Risk Management and Financing Disaster Losses in Developing Countries* (Karlsruhe, Germany: Verlag Versicherungswirtschaft GmbH).

Miguel, E., Gertler, P. & Levine, D. I. (2006) Does industrialization build or destroy social networks? *Economic Development and Cultural Change*, 54(2), pp. 287–317.

Mitchell, T., Haynes, K., Hall, N., Choong, W., & Oven, K. (2008) The roles of children and youth in communicating disaster risk, *Children, Youth and Environments*, 18(1), pp. 254–279.

Moody, J., & White, D. R. (2003) Structural cohesion and embeddedness: A hierarchical concept of social groups, *American Sociological Review*, 68(2003), pp. 103–127.

North, D. C. (1995) The new institutional economics and third world development, in: J. Harriss, J. Hunter, & C. M. Lewis (Eds) *The New Institutional Economics and Third World Development*, pp. 17– 26 (London: Routledge).

Okuyama, Y. (2003) Economics of natural disasters: A critical review. Paper presented at the 50th North American Meeting, Regional Science Association International, 20–22 November, Philadelphia.

Ostrom, E. (1990) *Governing the Commons. The Evolution of Institutions for Collective Action* (Cambridge: Cambridge University Press).

Park, C. (2006) Risk pooling between households and risk-coping measures in developing countries: Evidence from Bangladesh, *Economic Development and Cultural Change*, 54(2), pp. 423–457.

Rose, A. (2004) Defining and measuring economic resilience to disasters, *Disaster Prevention and Management*, 13(4), pp. 307–314.

Rose, A. (2007) Economic resilience to natural and man-made disasters: Multidisciplinary origins and contextual dimensions, *Environmental Hazards*, 7(4), pp. 383–398.

Santhanboa, C. (2008) Building houses in lampoon community after the Tsunami, in: Tsunami Aid Watch (Ed) *Building for the Future. A Communal Approach to Rehabilitation After the Tsunami*, pp. 8–18 (Chiang Mai, Thailand: Tsunami Aid Watch, Southeast Asia Regional Office).

Segschneider, K., & Worakul, W. (2007) *78 Weeks Later. A Descriptive, Quantitative and Qualitative Summary after the Tsunami in Thailand* (Chiang Mai, Thailand: Tsunami Aid Watch Programme, South East Asia Regional Office).

Sen, A. (1983) poor, relatively speaking, *Oxford Economic Papers*, 35(1983), pp. 153–169.

Sen, A. (1985) *Commodities and Capabilities* (Amsterdam: North-Holland).

Sharma, M., & Zeller, M. (1997) Repayment performance in group-based credit programs in Bangladesh: An empirical analysis, *World Development*, 25(10), pp. 1731–1742.

Siegel, P. B., & Alwang, J. (1999) *An Asset-Based Approach to Social Risk Management: A Conceptual Framework* Social Protection Discussion Paper Series No. 9926, Washington, DC: The World Bank.

Singh, N. C. (1996) *Community Adaptation and Sustainable Livelihoods: Basic Issues and Principles* (Winnipeg: International Institute for Sustainable Development).

Skoufias, E. (2003) Economic crises and natural disasters: Coping strategies and policy implications, *World Development*, 31(7), pp. 1087–1102.

Skoufias, E., & Parker, S. W. (2002) *Labor Market Shocks and their Impacts on Work and Schooling: Evidence from Urban Mexico*, IFPRI-FCND Discussion Paper No. 129.

Skoufias, E., & Quisumbing, A. R. (2004) Consumption insurance and vulnerability to poverty: A synthesis of the evidence from Bangladesh, Ethiopia, Mali, Mexico and Russia, *European Journal of Development Research*, 17(1), pp. 24–58.

Smith, K. (1992) *Environmental Hazards. Assessing Risk and Reducing Disaster* (New York: Routledge).

Sricharoen, T. (2006) Vulnerability and risk management for sustainable livelihoods of farm households in northern Thailand: The role of health insurance in managing risk. PhD thesis, Hohenheim University, 2006.

Swidler, A., & Watkins, S. C. (2009) 'Teach a man to fish': The sustainability doctrine and its social consequences, *World Development*, 37(7), pp. 1182–1196.

Syrett, S., & Evans, M. (2007) Generating social capital? The social economy and local economic development, *European Urban and Regional Studies*, 14(1), pp. 55–74.

Thomalla, F., Downing, T., Spanger-Siegfried, E., Han, G., & Rockström, J. (2006) Reducing hazard vulnerability: Towards a common approach between disaster risk reduction and climate adaptation, *Disasters*, 30(1), pp. 39–48.

Tompkins, E. L. (2005) Planning for climate change in small islands: Insights from national hurricane preparedness in the Cayman Islands, *Global Environmental Change*, 15(2005), pp. 139–149.

Tompkins, E. L., & Adger, W. N. (2004) Does adaptive management of natural resources enhance resilience to climate change? *Ecology and Science*, 9(2), pp. 10–23.

Trosper, R. L. (2002) Northwest coast indigenous institutions that supported resilience and sustainability, *Ecological Economics*, 41, pp. 329–344.

US White House (2006) *The Federal Response to Hurricane Katrina: Lessons Learned* (Washington, DC: US White House).

Wegerich, K. (2004) Informal network utilization and water distribution in two districts in Khorezm Province, Uzbekistan, *Local Environment*, 9(4), pp. 337–352.

Van Bastelaer, T., & Leathers, H. (2006) Trust and lending: Social capital and joint liability seed loans in southern Zambia, *World Development*, 34(10), pp. 1788–1807.

Zimmerman, F. J., & Carter, M. R. (2003) Asset smoothing, consumption smoothing and the reproduction of inequality under risk and subsistence constraints, *Journal of Development Economics*, 71, pp. 233–260.

Negotiation Support for Environmental Disaster Management: Drama Theory II and the Survival of Louisiana's Indigenous Gulf Coast Tribes

JASON K. LEVY
Homeland Security and Emergency Preparedness, L. Douglas Wilder School of Government, Virginia Commonwealth University, Richmond, VA, USA

ABSTRACT *A formal group decision and negotiation framework known as Drama Theory II (DT II) is put forth in order to systematically model environmental disasters and to examine ways to improve the resilience of Louisiana's indigenous Gulf Coast tribes that have lived on the bayous of the Mississippi delta for centuries. A mathematical treatment of key drama theoretic concepts (i.e. positions, intentions, doubts, and dilemmas) is provided, the dynamics of the drama theoretic process are discussed and theoretical results are applied to the ongoing oil spill disaster in the US Gulf Coast. While game theoretic tools have been widely applied to disaster management issues, traditional decision analysis assumes that decision-makers, options, and preferences are fixed. On the other hand, DT II, a theory of mutual interactive game design, allows for the redefinition of the conflict and emphasizes that decision-makers engage in a rational-emotional process of redefining both the game and their 'positions' in it until agreement on a satisfactory resolution is reached. Results show that drama theory constitutes an adaptive and powerful tool for modelling the Gulf Coast oil spill and for helping to strengthen the resilience of Louisiana's Gulf tribes which are on the brink of disaster. By identifying transformative natural resources policy options, it is shown how environmental health risks can be reduced and a more sustainable future can be promoted.*

Introduction

The twentieth century witnessed the development of powerful non-quantitative group decision and negotiation approaches for disaster decision-making and natural resources policy research including metagame theory (Howard, 1971). Drama Theory (DT) II constitutes a significant reformulation and extension of the original metagame theory and drama theory (Howard *et al.*, 1992). For more than 60 years, game theory has been widely used to analyze environmental disasters and crisis events (Osborne & Rubinstein, 1994). DT II was developed to deal with the shortcomings of traditional game theory. While drama theory takes an operations research approach to structuring, analyzing, and modelling complex problems, decision-makers, options, and preferences are not assumed to be fixed. DT II helps players interact to design the game they eventually play (by communicating with each other prior to playing a game) while simultaneously selecting a focal equilibrium set in that game. Specifically, DT II is valuable for analyzing large-scale 'wicked' (Rittel & Webber, 1973) disasters, including the uncertainties associated with predicting the consequences of a cascade of environmental, social, and economic impacts. For example, millions of gallons of crude have spilled into the sea due to the 2010 BP Gulf Coast oil spill but it is difficult to measure the long-term effects it will have on the marshlands and estuaries of Louisiana where fish and wildlife breed.

The 2010 Gulf Coast oil spill is modelled to better understand DT II and to provide critical insights on how to improve natural resources management and national energy policy. By adjusting strategy and deriving transformative policy options, it is shown how environmental disasters can be averted, energy needs can be met, and environmental risks can be reduced. DT II is an ideal tool for environmental disaster management because its theory of mutual interactive game design allows for conflict redefinition and emphasizes that decision-makers engage in a rational-emotional process of redefining both the game and their 'positions' in it until agreement on a satisfactory resolution is reached. Results show that drama theory constitutes an adaptive and powerful tool for modelling environmental disasters and for handling the inherently messy, complex, and challenging problems associated with them.

The paper is structured as follows. First, challenges facing the application of game theory to the modelling of environmental disasters are discussed. To overcome these perceived weaknesses, DT II concepts are put forth in order to identify the dilemmas facing decision-makers, to resolve these dilemmas, and to explore how the game should be redefined. Next, the similarities and differences of DT II and game theory are highlighted. Flowcharts for identifying dilemmas in Drama Theory II are provided and these dilemmas are applied to ongoing Gulf Coast oil spill with emphasis on natural resources policy implications. A thorough discussion and mathematical treatment of key drama-theoretic concepts (i.e. positions, intentions, doubts, and dilemmas) are provided. Finally, key conclusions are presented and future directions are put forth for improving environmental disaster decision-making.

Foundations of Drama Theory for Disaster Management

DT II assumes that decision-makers involved in managing a disaster engage in a type of pre-play drama-theoretic 'communication' (including words, explicit deeds, implicit acts, etc.) in order to achieve the 'common knowledge' necessary for responding to the disaster. Hence, drama-theorists employ a wider definition of rationality than the deeply embedded 'instrumental rationality' assumptions (i.e. static beliefs and preferences) of traditional game theory (Howard *et al.*, 1992; Howard, 1994, 1998). In contrast to game theory, drama theory asserts that managing disasters requires players to engage in a rational-emotional process of redefining both the game and their 'positions' in it until there emerges a satisfactory resolution on which they all agree. DT II involves activities that are not included in the narrower (traditional) game theoretic interpretation, including emotional appeals, apparent irrationality, and argumentation between parties (Rosenhead, 1989).

There are several key distinctions between drama theory and game theory with respect to disaster management. First, DT II makes the analysis more relevant and applicable to real-world disasters by eliminating the need for the cardinal (von-Neumann Morgenstern) payoffs and mixed (i.e. probabilistic) strategies used in game theory (drama theory uses ordinal preferences and pure strategies). Second, drama theory assumes any number of players, each of whom has a large number of options or alternatives to choose from. Third, game-theoretic analysis identifies equilibria within a fixed model which requires players to use artificially fixed preferences and options (and to make decisions based on moves in a model calibrated at the beginning of the analysis). This ignores the reality of disaster management, where time is spent debating how to avoid cascading impacts by analyzing dynamic and time-contingent strategies, preferences, and opportunities (Howard, 1998; Bryant, 2003). In contrast, DT II constitutes a theory of mutual interactive game design in which players, by communicating with others prior to managing the disaster, interact to design the game that they will ultimately play (including players, options, and beliefs).

Borrowing from theatrical terminology, an episode in DT II is defined in Figure 1 as moving through phases of *scene-setting* (which creates a closed informational environment) and *build-up* (which creates a common reference frame). In the *scene setting* stage, decision-makers collect data about the nature of the disaster and the relevant stakeholders in an Informationally Closed Environment (ICE). Next, decision-makers communicate *positions* (outcomes of the conflict that players would like to see realized) and *intentions* (the strategy that a player says she will carry out if current positions and stated intentions remain the same) together with *doubts* concerning each other's intentions. At this point, players share their beliefs and preferences about the unfolding disaster in which they face certain well-defined dilemmas. This is followed by the build-up phase, in which final positions

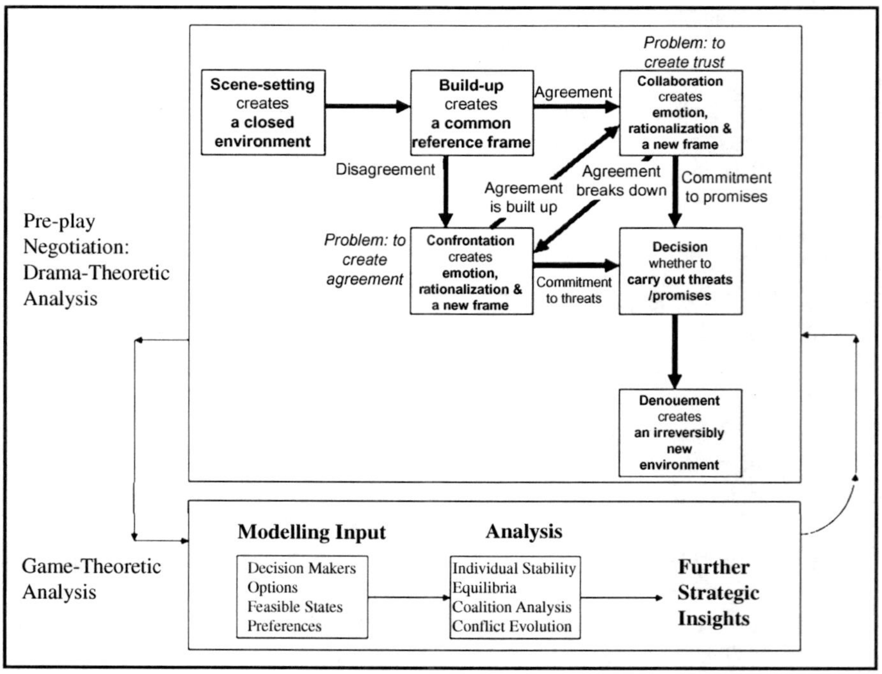

Figure 1. Relationship between DT II and game-theoretic analysis (Levy *et al.*, 2009).

are adopted within a common reference frame (CRF). During a disaster, decision-makers often exhibit mixed emotions, reaching a crescendo at the climax, after which players interact in terms of a redefined game. Following the build-up stage, if the players involved in managing the disaster are in agreement then collaboration ensues and efforts are focused on strengthening and confirming the agreement and solutions are devised to implement common positions.

On the other hand, if the stakeholder positions are not in agreement, the problem is one of overcoming conflict through the resolution of outstanding dilemmas, which arise when actors reach a moment of truth. The DT II framework assumes that dilemmas are created by each player openly stating a particular management solution to the disaster that it advocates (called its *position*) and communicating its intentions (the fall-back strategy in case the others reject its position). Taken together, stated intentions establish a threatened future if they contradict at least one decision-maker's position. Where positions disagree on the best way to manage a disaster, the result is confrontation and the parties adopt fallback strategies. Here, reason and emotion are used to change positions and fallback strategies. Parties will strive to eliminate dilemmas through rational debate and by establishing a new model. The process of dilemma-elimination takes place as positions and intentions are communicated during pre-play communication, before decisions are made. In analyzing preplay communication, DT II emphasizes the role of emotion and rational debate among stakeholders, two factors excluded from consideration in game-theoretic models. The stakeholders involved in any disaster management conflict are affected by strong emotions which trigger dilemma-eliminating rationalizations and redefinitions that in turn cause a change in crisis dynamics. Finally, the climax is followed by decision and denouement which establishes the next episode. At the decision phase, stakeholders make independent decisions to implement their disaster management positions or implement fallback strategies. During implementation, a new situation will arise as the players take irreversible actions.

Borrowing from theatrical terminology, DT II episodes are shown to move through phases of *scene-setting*, *build-up*, *climax,* and *decision* where the players may interact in terms of a redefined game. In drama theory terminology, an *episode* is a period of *pre-play communication* among decision-makers who, after communicating, engage in strategic interactions in a game defined through mutual dialogue. Following the *episode,* the game is played out, setting the stage for the next episode. Finally, the climax is followed by a moment of *decision* and *denouement*, the action that sets up the next episode. After a DT II analysis is complete the results can be input into a formal game theoretic analysis which uses information about players, options, and preferences to determine equilibria and strategic insights (Figure 1).

An Overview of the Three Dilemmas of DT II

A key objective in any drama theoretic analysis is to identify the dilemmas facing the decision-makers, each of which tends to cause emotions and rationalizations. This is achieved through the use of option boards, tables in which each party controls a number of 0/1 'options'. An options board (which can have any number of parties or options) is used to model the point in time in which the party has communicated its 'final' position (what it suggests each party should do) and 'final' stated intentions (what it says it will do in light of these positions and others' intentions). Hence DT II can help to model emotional responses that can lead to unanticipated reactions and change basic assumptions. By redefining this game and their positions in it, DT II allows players to eliminate their dilemmas in a large (non-Bayesian) world and provides disaster management professionals greater insights into the resolution of dilemmas. When all dilemmas are eliminated, the decision-makers agree on a solution that they do not doubt the other players will implement.

Dilemmas arise when players arrive at a moment of truth—this is when each player has taken up a position that it regards as final. These dilemmas cause emotions that decision-makers rationalize by producing new arguments as to why the game requires redefinition (Figure 2). If these arguments succeed, players potentially face new dilemmas within a new game. This continues until either they face no dilemmas or their arguments for a redefinition fail to convince. They then play the current game, in which they expect a certain equilibrium set. If and only if the process ends with none of these dilemmas, the players have resolved all their conflicts.

The three revised dilemmas in Drama Theory II are the persuasion, rejection, and trust dilemmas. Figure 2 identifies these three dilemmas for each option (O) and each party (P) and the immediate effects of each dilemma. The notation used in Figure 2 is also relevant for the remaining figures in the paper:

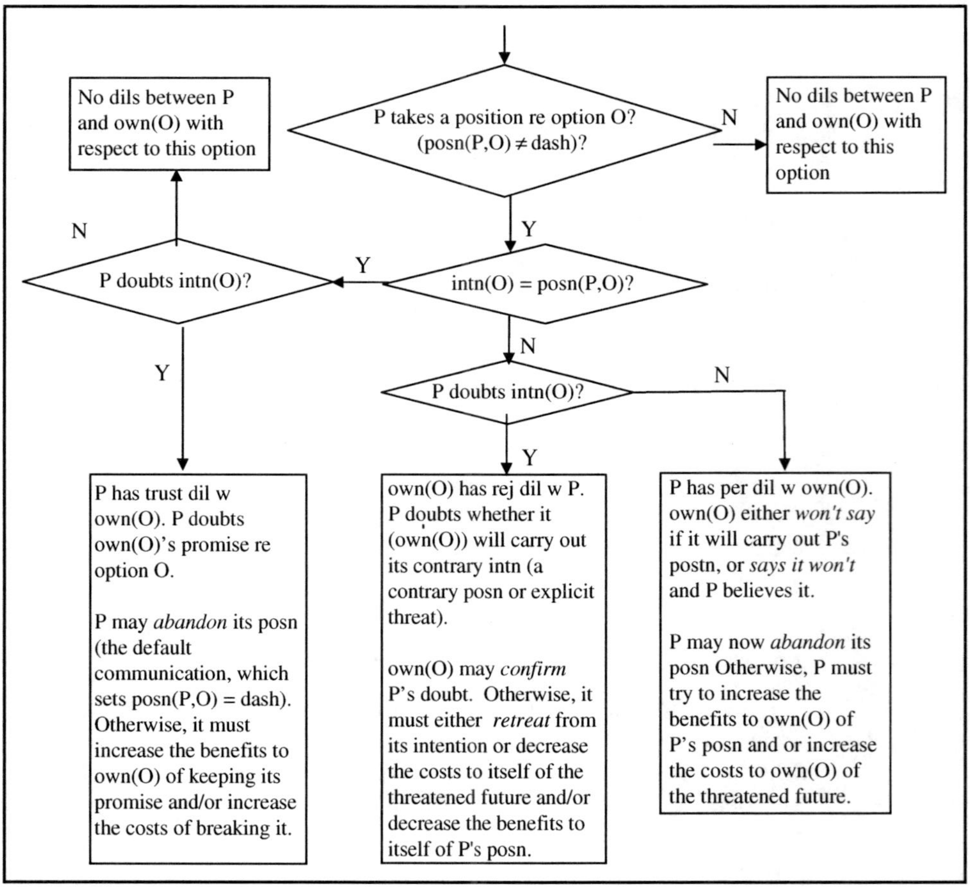

Figure 2. The identification of dilemmas for each option O and each party P (Levy *et al.*, 2009).

- O represents 'option'
- P represents 'player'
- posn represents 'position'
- intn represents 'intention'
- dils represents 'dilemmas'
- own(O) is the party that controls ('owns') option O. 'The owner of option O'
- intn(O) is 'adopted', 'rejected', or 'left open' (the last of these values being represented by a dash). It is own(O)'s stated intention for option O
- posn(P,O) is party P's position in relation to option O
- intn*(O) is the new, changed value of intn(O)
- posn*(P,O) is the new, changed value of posn(P,O)

The three dilemmas of drama theory are defined in relation to each option: A party may have a trust, rejection, or persuasion dilemma with any party in relation to any option owned by that party. The three dilemmas of DT II can be summarized as follows. Consider the situation in which actor A's intention contradicts actor B's position. If B has no doubts about A's intention, B has a persuasion dilemma with A over this option; however, if B does doubt it, then player A has a rejection/threat dilemma with B over this option. Hence the rejection/threat dilemma applies to options where player A intends to flout player B's position, and B doubts that A will. Finally, if A's intention is consistent with B's position for an option and B doubts it, then B has a trust dilemma with A over this option.

Let us now consider each dilemma in detail. The persuasion dilemma arises if either you won't say whether you will carry out my position, or you say you won't and I do not doubt it. More formally, player A has a persuasion dilemma with player B when A does not doubt B's intention to flout A's position (or threat). A implicitly 'gives in' to this dilemma if it communicates no new reasons why B should not flout its position (or threat); it 'fights' the dilemma by trying to communicate new reasons. Possible consequences of a persuasion dilemma are shown in the flowchart of Figure 3.

The trust dilemma can be simply defined as follows: I do not trust you to carry out your promise. Specifically, Player A has a trust dilemma with B if she doubts B's stated intention to meet A's position. A implicitly 'gives in' to this dilemma if she communicates no new reasons for B to adhere to her intention; she 'fights' by communicating new reasons. Possible responses to a trust dilemma are shown in Figure 4.

The final dilemma is the rejection dilemma: you do not believe that I will carry out a threat (or position) that conflicts with yours. Player A has a rejection dilemma to B when A's stated intention flouts B's position, but B doubts this intention. A implicitly 'gives in' to this dilemma if it communicates now new reasons why B should believe its intention. On the other hand, A fights the rejection dilemma by communicating new reasons why B should believe its intention. Possible responses to a trust dilemma are shown in Figure 5.

Defining Positions, Intentions, and Doubts

Consider a general game G in which we assume, without loss of generality, that the set of actions available at one information set does not intersect with the set available at another. Let the game G have:

- a set I of players (a set that does not include chance, although chance may play a role in the game);

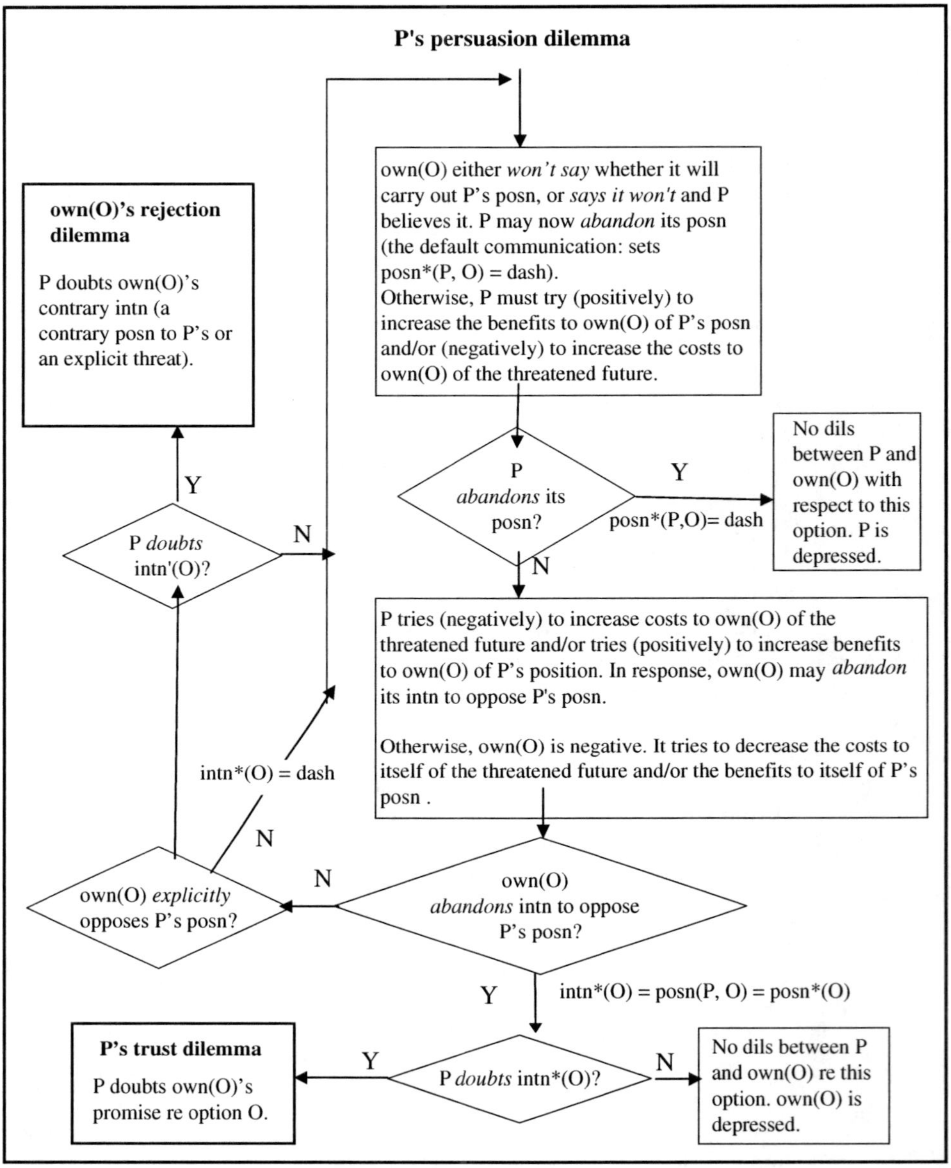

Figure 3. Possible consequences of a persuasion dilemma (Levy *et al.*, 2009).

- a set A of actions open to the players in I;
- a partition H of the set A such that each element of H represents an information set (a representation made possible by the above assumption of non-overlapping information sets);
- a player function P: $H \rightarrow I$ giving the player $P(h)$ that chooses an action from information set h.

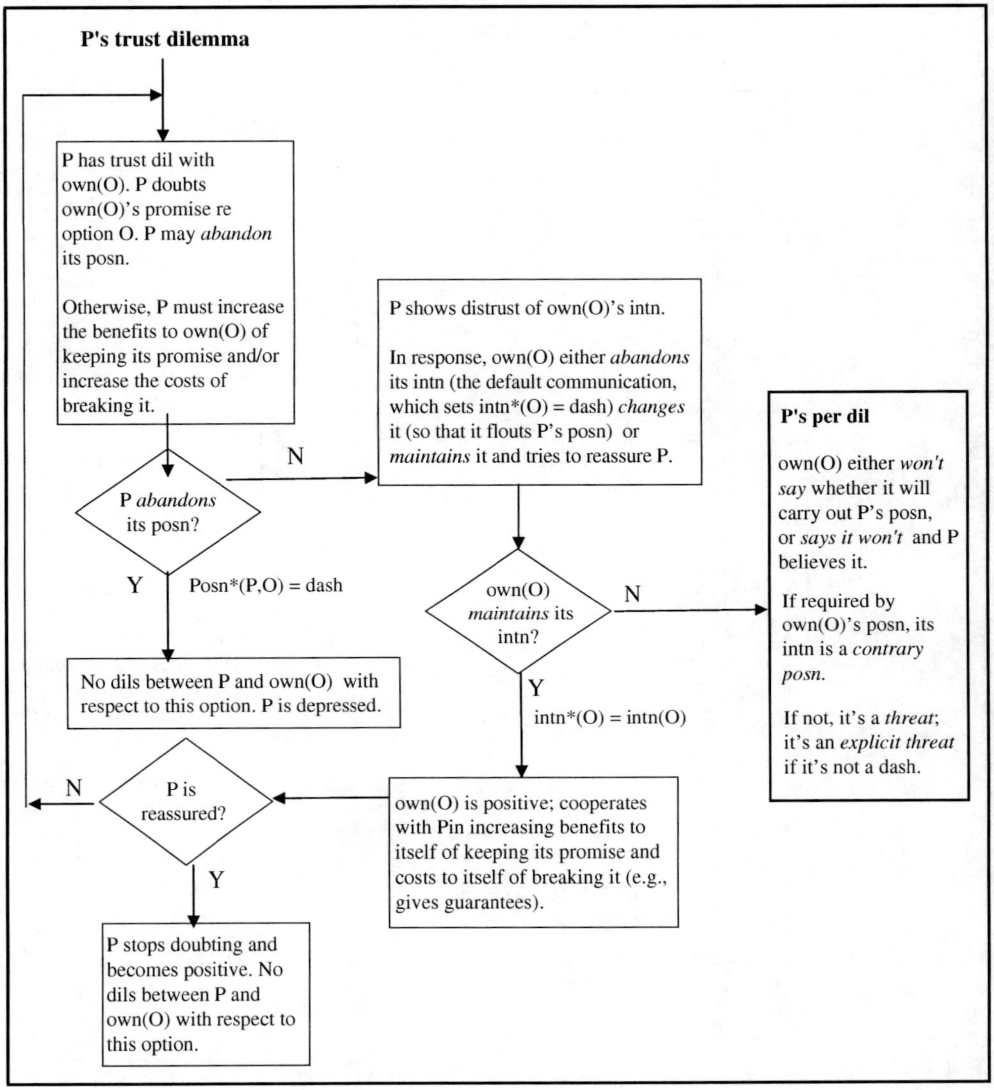

Figure 4. Possible responses to a trust dilemma (Levy *et al.*, 2009).

Positions, intentions, doubts, and dilemmas are defined as follows:

For each player $i \in I$, i's *position* p_i is a subset of A such that for all $h \in H$ we have p_i h $\neq \emptyset$; that is, i's position intersects with every information set.

- Interpretation: *pi* is the set of actions *i* does not object to (does not demand should have zero probability);
- Conversely, $A \setminus pi$ is the set *i* objects to (demands should have zero probability).

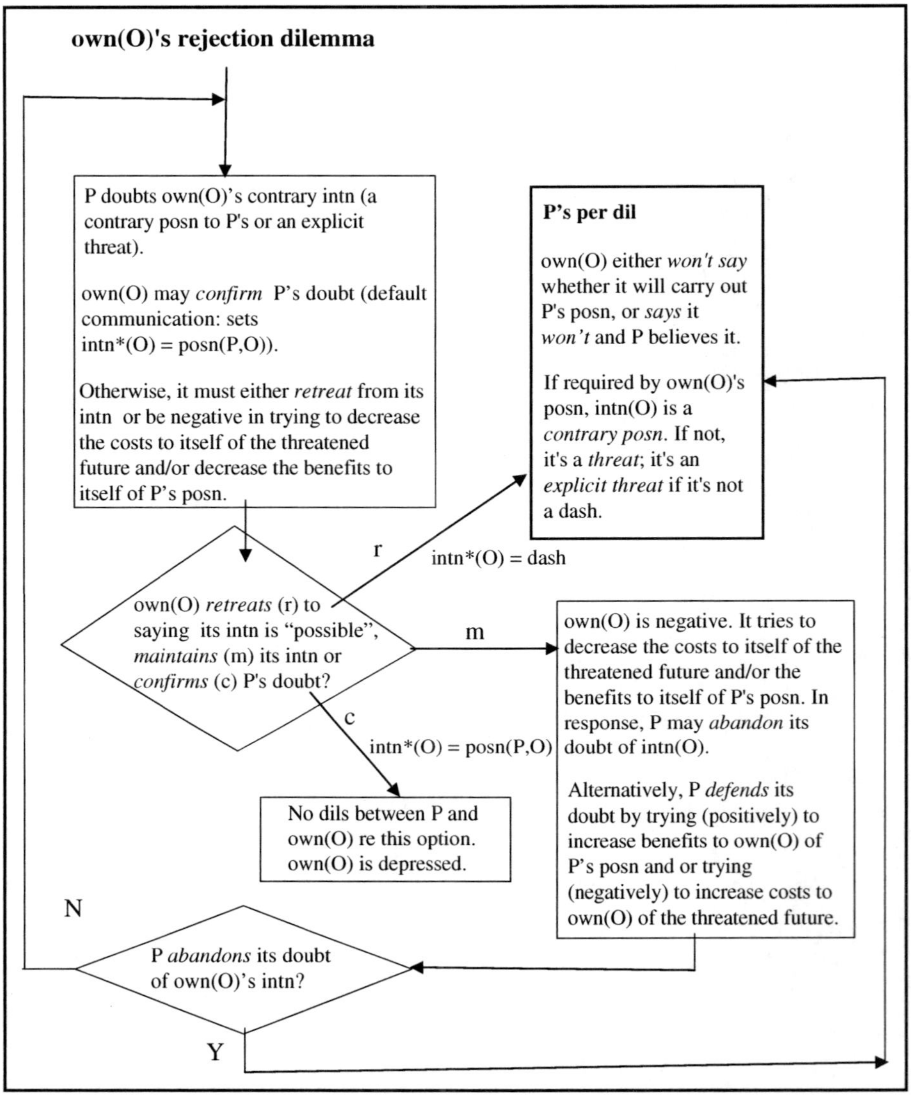

Figure 5. Possible consequences of a rejection dilemma (Levy *et al.*, 2009).

For each $i \in I$, i's *intentions* n_i are a subset of Ti (the set of player i's actions—i.e. the union of all i's information sets) such that for all h with $P(h) = i$ we have $n_i \quad h \neq \emptyset$; that is, i's intentions intersect with every i-owned information set.

- Interpretation: n_i is the set of i-actions that i claims to be considering (i.e. claims not to have assigned zero probability to);
- Conversely, $Ti \setminus n_i$ is the set i claims to have excluded (assigned zero probability to).

For each $i \in I$, i's *doubts* (or *doubted actions*) are a subset di of A such that for all h we have $di \;\; h \neq n \;\; h$; that is, i's set of doubted actions within an information set must not coincide with the stated intentions in that set. This is because if all intentions (actions said to be under consideration) in an information set are doubted by i, then some action said to be excluded must also be doubted; otherwise, i would be assuming that all actions in the information set had zero probability. Interpretation: d_i is the set of actions a such that, unless current rationalizations are rebutted, the claim concerning a is seen as false by i.

Definition and Interpretations of Dilemmas

The above definitions allow for the discussion of three dilemmas by defining three sets of dilemmatic actions. The following three sets specify player i's *persuasion, rejection,* and *trust dilemma* with respect to player j.

$$Per(i,j) = nj \backslash di \backslash pi.$$

$$Rej(i,j) = ni \cap dj \backslash pj.$$

$$Tru(i,j) = Tj \cap di \backslash nj \backslash pi.$$

i's *persuasion dilemma* with j is the set of j-actions that i objects to (they are not in p_i) and j claims to be considering (they are in n_j); moreover, j's claim is not doubted by i (none of the actions is in d_i). i must either try to persuade j to exclude them or abandon its position; that is, cease to object.

i's *rejection dilemma* with j is the set of i-actions that j objects to (they are not in p_j) and i claims to be considering (they are in n_i)—but its claim is doubted by i (they are in d_i). i must either try to persuade j it does not exclude them or admit that it does. The latter is the default communication—i.e. by not trying to persuade j that it does not exclude such an action, i acknowledges that it does.

i's *trust dilemma* with j is the set of j-actions that i objects to (they are not in p_i) and j claims to exclude (they are not in n_j)—but its whole claim is doubted by i (they are all in d_i). i must either find ways to ensure that j can be trusted or abandon its position. The latter is the default communication—i.e. by not trying to find ways to ensure that j excludes such an action, i acknowledges that it does not object to it.

DT II Disaster Management Case Study

The following disaster management case study is modelled using concepts from DT II with the assistance of the Decision Support System Confrontation Manager^{TM} (which supports a previous version of drama theory). Option board diagrams were obtained using Windows Paint^{TM} to adjust output from Confrontation Manager^{TM}. Our focus is on protecting the civilization of Louisiana's indigenous Gulf Coast tribes from the recent 2010 Horizon oil spill. Our initial disaster model contains three actual decision-makers, namely Louisiana governor Bobby Jindal (LA), the US Army Corps of Engineers (ACE), and the US Federal Government (Fed). Note that the US Federal Government has two options it controls while all the other parties have one (offshore drilling moratorium and expand

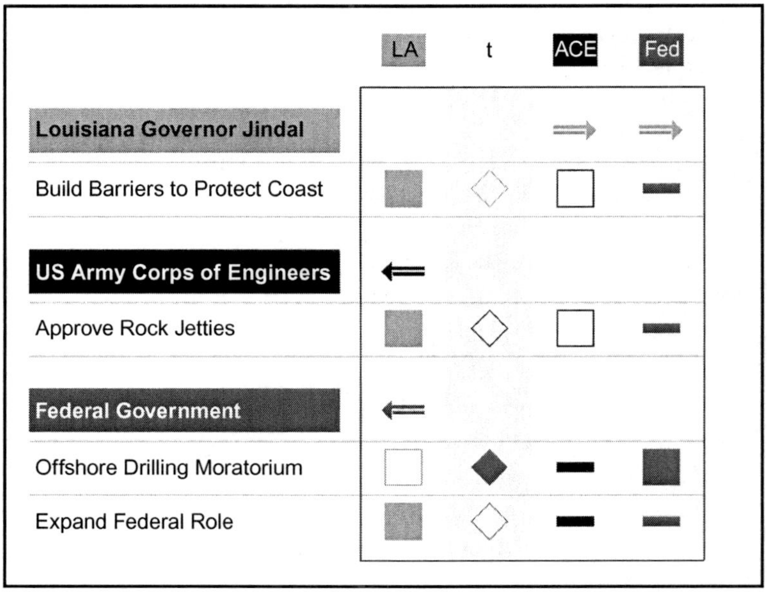

Figure 6. Oil spill encounter options board.

federal role) (Figure 6). The preferences, options, and intentions of these three key stakeholders are discussed together with the dilemmas facing all players and the short and long term implications of each dilemma. The options in Figure 6 constitute the minimum needed to specify parties' positions and stated intentions. Acceptance or rejection of an option is always represented by a filled-in or empty shape.

If it is uncertain, conditional, or left open whether an option should or will be adopted or rejected, a dash is inserted. In Figure 6, the first column shows the position of Governor Jindal (LA). He proposes to build barriers to protect the Louisiana coast (solid square in row 1), and would like the US Army Corps of Engineers to quickly approve rock jetties (solid square in row 2). His position also includes ending the offshore drilling moratorium (open square in row 3) and an expanded federal role in managing the Gulf Coast disaster (solid square in row 4).

Louisiana Governor Bobby Jindal has criticized US federal officials over delays in constructing sand berms off the Chandeleur Islands to block channels between these barrier islands. Specifically, Jindal supports Jefferson Parish's request to build a rock barrier in order to protect oil from reaching sensitive regions such as Barataria Bay, where many of the smaller islands and habitats have become soiled. In June 2010, Jefferson Parish requested that the rock dikes be constructed for five passes along Barataria Bay although the parish later reduced the number of rock barriers in the plan down to two (both east of Grand Terre near Grand Isle): Four Bayou Pass and Pass Abel. Governor Jindal has also clashed with the federal government over the need to rapidly send additional oil skimmers and protective boom to coastal Louisiana.

The second column illustrates the US Army Corps of Engineers' (ACE) position. ACE is reluctant to build barriers and approve rock jetties (and does not take a position on the remaining US government options). Specifically, the US Army Corps of Engineers rejected

Jefferson Parish's plan to use rock jetties (jetties) near Grand Isle to stop oil from reaching deep into Barataria Bay, an ecologically rich region. In addition, the Environmental Protection Agency (EPA), the National Oceanic and Atmospheric Administration (NOAA), and Louisiana's scientific review panel (the Horizon Oil Spill Science and Engineering Review Team) have expressed numerous concerns with the rock jetty plan. They fear that the rock dikes could damage coastal resources (more than the oil) since blocking the flow of water through the passes will increase the water velocity which will potentially deepen the channel and destroy natural 'soft' structures such as existing sand islands and marshes. 'The softer parts of the system (e.g. the sandy barriers between the inlets) will then become the weak spot as the inlets have hardened and constricted', noted Denise Reed, a University of New Orleans coastal scientist. She concluded that 'hardening the inlets makes breaching of the islands more likely—both resulting in additional erosion and more pathways for oil to move in from the Gulf' (Kirkham, 2010). Echoing these concerns, a letter from a separate group of 15 coastal scientists to the US Army Corps of Engineers stated that more scientific and technical expertise from environmental groups and outside experts is needed to improve the planning process and strategies to reduce the impact of the oil spill.

The US Army Corps of Engineers and other groups are also worried that there are insufficient plans for removing the rock dikes once the oil threat eventually diminishes. In the 10 July 2010 letter denying the permit, Colonel Alvin Lee, the commander of the US Army Corps of Engineers' New Orleans District, noted that baseline data from which to assess project-related damages was insufficient and that a detailed plan of action (agreed to by all relevant parties) was missing. Colonel Lee explicitly mentions his concerns that remedial actions could harm the environment since 'no responsible party to ensure timely structure removal to minimize environmental harm has been identified; no restoration plan to mitigate environmental damages has been furnished' (Kirkham, 2010).

Responding to these concerns Governor Jindal notes that Jefferson Parish's revised rock barrier plan will include an escrow account to be funded by BP PLC in the event that the rock jetties must be removed due to environmental damage. The Shaw Group, the proposed contractor for the rock barriers is developing a monitoring plan capable of identifying morphological changes to the barrier islands and passes: 'Should serious unexpected morphological changes be observed, the proposed rock structures will be altered or removed to correct the problem' (Kirkham, 2010). Governor Jindal is gravely concerned that the government has not proposed an alternative plan to the proposed rock dikes.

The fourth column illustrates the position of the federal government. They initially do not take a position on any option except for promoting a deepwater offshore drilling moratorium. While National Incident Commander Thad Allen and his unified command organization have carried out excellent response work under difficult conditions, many have called for President Obama and his administration to play a larger role in managing the disaster. For example, the Secretary of Homeland Security could play a larger leadership role under Homeland Security Presidential Directive-5 (HSPD-5) as the 'principal Federal official for domestic incident management' by 'coordinating involvement of senior Administration officials in joint oil/hurricane response planning for the region; addressing the State Department's failure to effectively coordinate offers of international assistance; and, working with the Secretary of the Navy to execute his new Gulf Coast recovery responsibilities' (Paulison & Kaniewski, 2010). Moreover, interagency collaboration could be enhanced by employing the National Response Framework (NRF) which is the only approach for managing a natural disaster together with the oil spill.

The intentions of each player are shown in the second column (entitled *t* for threatened future). If all positions remain as shown in Figure 6 the result will be that no protective barriers will be constructed, a deepwater offshore drilling moratorium will be implemented, and the federal government will not play a larger role. One way to improve the situation is to include the needs of a new player: the First Nations Tribes in the Gulf Coast region. For example, the livelihood and culture of the Houma First Nations Indian Tribe are under tremendous pressure.

Policy Implications

A symbiotic relationship exists between Indian tribes in the US state of Louisiana and the natural environment. For example, the United Houma Nation, a state-recognized tribe whose 17 000 members live in six parishes in the US state of Louisiana, work the waters of Bayou Pointe–Aux-Chenes and its nearby bays and lakes for shrimp, fish, crabs, oysters, and crawfish. While the Houma have overcome disease and hurricanes, 'nothing of this magnitude has ever hit the Gulf Coast' and the indigenous tribes are concerned about the survival of their civilization (Weinberg, 2010). The Gulf Coast oil spill has the potential to destroy the livelihoods, landscape, and cultural identity of the United Houma Nation and other First Nations people living off the land and water in the bayous of southern Louisiana. The oil spill has already ruined oyster plots, soiled crab traps, and cut off shrimp trawlers from some of this area's best fishing grounds (Goodman, 2010). As declared by Principal Chief Brenda Dardar Robichaux last week in her testimony before members of the House Subcommittee on Insular Affairs Wildlife and Oceans, which is investigating the impacts of the BP spill: 'This lifestyle is now in jeopardy' (Robichaux, 2010).

There are a number of forces threatening the survival of the Houma culture. First, canals dug for oil and gas exploration since the 1930s have caused the deterioration of wetlands: saltwater intrusion kills the marsh grasses (and storms wash away the soil left behind) while levees keep the nearby Mississippi River from replenishing the marshland's soil. Second, oil and gas extraction has led to land subsidence in some places. As a result, the Mississippi delta is shrinking: water now fills areas that were once freshwater marshes. For decades, the Houma and other native groups here have adapted to the loss of wetlands and the encroachment of the oil and gas industry; however, the short- and long-term impacts of the oil spill may force the community to relocate. Third, the oil and dispersants have the potential to devastate the fisheries and natural environment on which Louisiana's Gulf Coast tribes depend. For hundreds of years, the Houma and other Louisiana tribes have been hunting, fishing, shrimping, crabbing, trapping, and harvesting oysters in Louisiana's coastal marshes and wetlands.

Finally, the health effects of the oil and dispersants could adversely impact cleanup workers and communities. In particular, Houma Chief Robichaux emphasized the need to improve training of cleanup workers.

> because people who are actually doing some of the cleanup work, we feel, are not really being educated on the long-term impacts of what this is going to do to their health. Normally, the training that is done lasts 40 hours . . . if you are going to work with this hazardous material. The training that they are being given is actually a four-hour training. (Robichaux, 2010)

The 34 000 workers that are cleaning up the Gulf Coast spill—the largest oil spill in US history—must undertake hazardous tasks, in low boats close to oil on the water (while chemical dispersants are sprayed overhead). Breathing in or swallowing oil and/or dispersant can cause nausea, vomiting, and possibly throat and lung irritation. Crude oil has a complex chemical profile including carcinogenic compounds that can alter a human's DNA. The chemistry of the oil constantly evolves as it ages and moves from the open ocean to the shoreline. 'Fresh' crude contains volatile organic compounds such as the carcinogen benzene, known to cause cancer in humans while polycyclic aromatic hydrocarbons and other substances can poison the brain and central nervous system. More rigorous training and supervision could help to protect workers, especially volunteers (Stier, 2010).

BP and US government officials are involved in one of the largest and most aggressive experiments involving chemical dispersants in the world (Rosenthal, 2010). The effects of using these large amounts of dispersants are unknown: dispersants have been sprayed on the Gulf of Mexico's surface and are pumped directly onto the leak (a mile beneath the surface) in order to break up the oil into smaller droplets (that more easily biodegrade). While BP PLC reduced its use of Corexit in response to a federal directive, BP has still deployed nearly 2 million gallons of dispersant in the Gulf of Mexico since April 2010 in its effort to contain the Deepwater Horizon oil spill, sparking concerns among scientists, environmental groups, public-health advocates, and politicians. The result is a vast expanse of oil droplets mixed with dispersants in the Gulf of Mexico. The EPA has yet to examine the toxicity of dispersants when mixed with crude oil. EPA Administrator Lisa Jackson has stated that she is 'unaware of any research' on the effects of applying the dispersants in the unprecedented volume that has entered the Gulf region (Schor, 2010).

The immense scale of the deployment of dispersants in the Gulf Coast is unprecedented (approximately one-third of the world supply), raising several key issues about the ultimate fate of the chemically dispersed oil and the long-term effects on organisms exposed to it. For example, BP released large quantities of dispersants from the air on miles of oil slicks across the Gulf, yet it is not known whether dispersants are distributing the oil to a more widespread area—or increasing its toxicity (oil is toxic at 11ppm, Corexit 9500 at only 2.61ppm). While dispersants enhance biodegradation in the aquatic environment, important questions include: What is the effectiveness of dispersants under various environmental conditions? How will the different oil exposure pathways due to dispersant use affect sea life? What are the effects of using dispersants in the deepwater (where cold temperatures and high pressure deep below the surface prevent the oil from breaking down)? Is the widespread use of dispersants shifting the environmental impacts from the coastal environment (ecosystems along the shore) to marine ecosystems (i.e. fisheries and fishermen)?

Neither BP nor the US Environmental Protection Agency (EPA) has found an effective alternative to Corexit and BP refused a government order to find a less toxic alternative to the Corexit line of dispersants. The UK's Marine Management Organization does not allow the use of Corexit for rocky shoreline application because the dispersant was found harmful to sea life (i.e. limpets, snails and other crustaceans) that inhabit rocky shores. And while most scientists support the application of dispersants as a necessary evil to limit the accumulation of dense oil on the surface or washing up onshore (places where it is most likely to harm coastal marine life), there is concern about filling the Gulf of Mexico with low concentrations of dispersants (Rosenthal, 2010).

The CDC, along with state and local health departments in the affected Gulf Coast states, is conducting surveillance for short-term health effects possibly related to the oil spill using national and state-based surveillance systems (including the worsening of asthma, cough, chest pain, eye irritation, nausea, and headache). The Centers for Disease Control (CDC) notes that 'if possible, everyone, including pregnant women, should avoid the oil and spill-affected areas' (CDC, 2010) and recommends that cleanup workers use a N95 respirator mask. There is a dearth of short- and long-term research examining the human health effects of oil exposure: only 7 of 38 supertanker oil accidents have been studied for health impacts (Aguilera *et al.*, 2010). According to safety documents posted by Nalco, the Corexit dispersants should not touch the skin and can damage lungs (Nalco, 2010). By routinely monitoring oil cleanup workers and tracking symptoms (especially those related to the eyes, skin, and respiratory, cardiovascular, gastrointestinal, and neurological systems) it should be possible to detect early signs of problems (Wechsler & Gross, 2010).

While BP claims it is working with federal health and environmental regulators to protect cleanup workers from the hazards of the Gulf spill, experts are concerned that the workers may suffer similar illnesses to those suffered in previous spills. For example, Rikki Ott, an environmental activist who studied the 1989 Exxon Valdez spill in Prince William Sound, Alaska, said cleanup workers in the Gulf are showing 'the exact identical symptoms down here that we had 21 years ago' (Griffin, 2010). The EPA revealed that Corexit 9527 has been applied in the Gulf spill. This formula contains 2-butoxyethanol 'the ingredient known to have caused health problems with the workers of the Exxon Valdez spill' (Safer States, 2010). Dr. Blanca Laffon, a genetic toxicologist and public health researcher at the University of Coruña, has carried out studies on the long-term ramifications of exposure to crude oil toxins. Her team took blood samples from and studied biomarkers in two groups of workers that were exposed to the 2002 Pestige Oil spill (in which the Prestige oil tanker sank off Spain's northwestern coast and polluted thousands of miles of Spanish and French coastline). The first group was composed of volunteers exposed to oil for only five days (while handling soiled birds); while the second group was comprised of cleanup workers hired by the government for several months (to remove oil from rocks and beaches). An increase in DNA damage (and hence a heightened risk to cancer) was found in both groups with the amount of damage related to the amount of time they were exposed to oil. In the long run, DNA damage was repaired in the group exposed to oil for only five days, while the DNA damage became fixed in chromosomal DNA in workers exposed for months (Pérez-Cadahía *et al.*, 2008; Laffon *et al.*, 2006; Mascarelli, 2010).

According to Irwin Redlener, a member of the National Commission on Children and Disasters and director of Columbia University's National Center for Disaster Preparedness, Gulf Coast children are among the most vulnerable population—after the spill workers—for a number of reasons. First, children are at a higher risk than adults because their bodies detoxify chemicals less efficiently. Second, their shorter stature means that they are may also be more likely to breathe in heavier-than-air gas pollution found near the ground. Third, children may also be unaware or unable to understand public health guidelines and put a variety of substances in their mouths. Fourth, children are more prone to absorption through their skin, inhalation problems, and ingestion. Finally, children in the region are still suffering psychologically and physically from the combined effects of the 2005 Gulf Coast hurricanes, Katrina, and the oil spill (Wechsler & Gross, 2010; Stier, 2010).

It is essential to improve chemical testing in order to gather consistent and clear data on dispersant toxicity and track the extent of the exposure and to improve oil spill response judgments under extreme uncertainty. While the impacts of the dispersants are not well understood, low concentrations of Corexit may be toxic to marine life, including crustaceans, algae, and fish larvae (Pitman, 2010). In response to concerns about the limited availability of test data, US EPA Administrator Jackson has called for more testing and disclosure of the chemicals found in dispersants with a focus on monitoring the chemicals' long-term effects (on human and marine ecosystem health) and ingredient reporting (Schor, 2010). Many politicians, scientists, and environmental groups have assailed the ability of chemical manufacturers to resist releasing the ingredients and formulas for chemical products by filing confidentiality claims since keeping the chemical composition mix hidden makes it more difficult to evaluate the risks. Specifically, EPA Administrator Jackson has asked Congress to pass legislation bolstering the EPA's authority over dispersants and reforming the 1976 Toxic Substances Control Act (TSCA), the federal law that determines what toxic chemicals humans are exposed to—and shields the ingredients in chemical products (Schor, 2010). There are bill proposals currently in the US Senate (The Safe Chemicals Act of 2010) and the US House (The Toxic Chemicals Safety Act) which, if passed, will begin to overhaul the TSCA.

A number of field investigations have been carried out to evaluate the toxic effects of crude oil and types of Corexit currently being applied to the Gulf of Mexico on aquatic species. However, laboratory toxicology results are highly dependent on aquatic test species and experimental conditions which may not be 'representative of actual exposure conditions' (George-Ares & Clark, 2000). Hence, oil spill response coordinators should not base their decisions related to the use of dispersants solely on short-duration laboratory aquatic toxicology results. Moreover, higher Gulf Coast water surface temperatures may increase the toxicity of Corexit chemicals.

The Louisiana state health department's oil spill surveillance report notes that 128 cleanup workers (on oil rigs or individuals involved in the oil spill cleanup efforts) were believed to have been sickened due to exposure to the Gulf of Mexico oil spill by the end of June 2010—in addition to 34 cases of illness affecting the general public (CNN, 2010). Of these 162 cases, the most commonly reported symptoms were throat irritation, dizziness, shortness of breath, cough, eye irritation, nausea, and headaches. At least 21 people were briefly hospitalized when they developed symptoms.

A surveillance network of doctors, clinics, emergency care locations, and medical facilities have concluded that health problems from the Gulf spill are not limited to physical ailments and illnesses and may extend to a potential mental health crisis in communities affected by the oil spill. Specifically, health teams counselling families and communities affected by the spill have reported a spike in the demand for mental health services due to reported increases in depression, stress, grief, and alcoholism. In a letter to BP Chief Operating Officer Doug Suttles requesting funding for health services, Louisiana Health and Hospitals Secretary Alan Levine warns of 'an emerging behavioural health crisis... There exists anger, anxiety and uncertainty among the families and communities affected by the spill, which will easily manifest into addiction and various forms of mental health crisis if not confronted' (CNN, 2010). In response to the Gulf Coast spill, Louisiana has deployed a full suite of mental health services: outreach teams have counselled almost 2000 people in affected areas.

In order to cope with the spiralling web of mental and behavioural health consequences arising from Hurricane Katrina, outreach teams from Louisiana Spirit (a subdivision of the Louisiana Department of Health and Hospitals—Office of Mental Health) provided counselling to hurricane victims. Originally funded by the Federal Emergency Management Agency (FEMA) and the federal Substance Abuse and Mental Health Services Administration (SAMHSA), after the Gulf oil spill Louisiana Spirit has returned. The programme is currently financed by a US$1 million grant by BP to Louisiana (slated to last until August, 2010). US National Institutes of Health (NIH) Director Francis Collins, has announced that the NIH has allocated US$10 million to study the public health impact of the oil spill and will focus on cleanup workers and residents of the Gulf Coast. According to Alan Levine, many Gulf Coast victims are exhibiting the 'early warning signs of developing substance abuse and dependence, mental illness, suicide and familial breakdown including divorce, spouse abuse and child abuse and neglect' (CNN, 2010). As a result, clinical treatment for stress disorders and other mental illnesses may be necessary and Louisiana government officials are calling on BP to fund mental health programmes.

Conclusions

DT II, a theory of pre-game communication and equilibrium selection, allows for the modelling of time-sensitive, multi-faceted, and multiple participant disaster negotiations. While game theoretic tools have been widely applied to natural resources and disaster issues, traditional game theory assumes that decision-makers, options, and preferences are fixed. DT II was shown to constitute an effective tool for managing the Gulf Coast oil spill and improving the resilience of Louisiana's indigenous gulf tribes. The general substantive conclusion from our modelling is that the survival of Louisiana's Gulf Coast tribes depends on improving the natural environment and working closely with local, state, and federal agencies. With foundations in game theory, drama theory provides a valuable alternative to game-theoretic approaches to resolving conflict by examining the emotions and rationalizations that play a role in reducing conflict and eliminating key dilemmas. DT II is a decision technology which requires only relative (ordinal) preference information and is highly appropriate in coping with 'unmeasurable' disaster impacts. In summary, drama theory constitutes a theory of pre-game communication which allows would-be players to establish and redefine the game they ultimately play. A 'fictitious' party called Assumptions (A) could also be used to attach consequences (i.e. extinction of First Nations' cultures) to the intentions or positions of all parties. Finally, this work can be extended for global oil spill management by building on experimental results which have illustrated the validity of DT II for a number of natural, heath-related, and technologic threats.

References

Aguilera, F., Méndez, J., Pásaro, E., & Laffon, B. J. (2010) Review on the effects of exposure to spilled oils on human health, *Journal of Applied Toxicology*, 30, pp. 291–301.

Bryant, J. W. (2003) *The Six Dilemmas of Collaboration* (Chichester, UK: Wiley).

CDC (2010) Gulf oil spill information for pregnant women. Emergency preparedness and response. Available at http://www.bt.cdc.gov/gulfoilspill2010/2010gulfoilspill/pregnancy_oilspill.asp (accessed 1 September 2010).

CNN (2010) 162 cases of illness linked to oil spill reported in Louisiana, 28 June. Available at http://www.cnn.com/2010/HEALTH/06/28/louisiana.oil.exposure/?iref=obnetwork.

George-Ares, A., & Clark, J. R. (2000) *Acute Aquatic Toxicity of Three Corexit Products: An Overview*, prepared for Exxon Biomedical Sciences, Inc. Available at www.iosc.org/papers/00020.pdf (accessed 2 September 2010).

Goodman, J. (2010) Gulf oil spill threatens tribe's livelihood in Louisiana, *Miami Herald*, 1 June. Available at http://www.miamiherald.com/2010/05/31/1657504/united-houma-nation-faces-biggest.html.

Griffin, D. (2010) Critics call Valdez cleanup a warning for Gulf workers, *CNN*, July 7. Available at http://www.cnn.com/2010/US/07/07/oil.spill.valdez.workers/index.html?hpt=T2.

Howard, N. (1971) *Paradoxes of Rationality: Games, Metagames and Political Behavior* (Cambridge, MA: MIT Press).

Howard, N. (1994) Drama theory and its relation to game theory. Part 2: Formal model of the resolution process, *Group Decision and Negotiation*, 3, pp. 207–235.

Howard, N. (1998) N-person 'soft' games, *Journal of Operational Research Society*, 49, pp. 144–150.

Howard, N., Bennett, P. G., Bryant, J. W., & Bradley. M. (1992) Manifesto for a theory of drama and irrational choice, *Journal of Operational Research Society*, 44, pp. 99–103.

Kirkham, C. (2010) Battle over plan to set up rock dikes looks like replay, *Times Picayune*. Available at http://www.nola.com/news/t-p/frontpage/index.ssf?/base/news-15/1278484831206900.xml&coll=1 (accessed 7 July 2010).

Laffon, B., Fraga-Iriso, R., Pérez-Cadahía, B., & Méndez, J. (2006) Genotoxicity associated to exposure to Prestige oil during autopsies and cleaning of oil-contaminated birds, *Food and Chemical Toxicology*, 44, pp. 1714–1723.

Levy, J. K. Hipel, K. W., & Howard, N. (2009) Advances in drama theory for managing global hazards and disasters. Part I: Theoretical foundation, Part II: Coping with global climate change and environmental catastrophe, *Group Decision and Negotiation*, 18(4), pp. 303–334.

Liua, B., Romairea, R. P., Delauneb, R. D., & Lindaub, C. W. (2006) Field investigation on the toxicity of Alaska North Slope crude oil (ANSC) and dispersed ANSC crude to Gulf killifish, Eastern oyster and white shrimp, *Chemosphere*, 62(4), pp. 520–526.

Mascarelli, A. (2010) Oil-spill health risks under scrutiny, *Nature News*, 24 June. DOI:10.1038/news.2010.316. Available at http://www.nature.com/news/2010/100624/full/news.2010.316.html.

Nalco Company (2010) Nalco oil dispersant information, 18 July. Available at http://www.nalco.com/news-and-events/nalco-oil-dispersant-information.htm.

Osborne, M. J., & Rubinstein, A. (1994) *A Course in Game Theory* (New York: MIT Press).

Paulison, R. D., & Kaniewski, D. (2010) The Gulf oil disaster: Three steps to federal leadership (HSPI website). Available at http://www.gwumc.edu/hspi/policy/commentary14_OilSpill.cfm (accessed 3 September 2010).

Pérez-Cadahía, B., Méndez, J., Pásaro, E., Lafuente, A., Cabaleiro, T., & Laffon, B. (2008) Biomonitoring of human exposure to prestige oil: effects on DNA and endocrine parameters, *Environmental Health Insights*, 2, pp. 83–92.

Pitman, C. (2010) Despite concerns, dispersant use continues on Gulf oil spill, *St. Petersburg Times*, 25 June. Available at http://www.tampabay.com/news/environment/water/despite-concerns-dispersant-use-continues-on-gulf-oil-spill/1104771.

Rittel, H., & M. Webber (1973) Dilemmas in a general theory of planning, *Policy Sciences*, 4, pp. 155–169.

Robichaux, B. D. (2010) House Subcommittee on Insular Affairs Wildlife and Oceans. Available at http://resourcescommittee.house.gov/index.php?option=com_frontpage&Itemid=83 (accessed 3 September 2010).

Rosenhead, J. (Ed) (1989) *Rational Analysis for a Problematic World* (New York: Wiley).

Rosenthal, E. (2010) In Gulf of Mexico, chemicals under scrutiny, *New York Times*, 5 May. Available at http://www.nytimes.com/2010/05/06/science/earth/06dispersants.html.

Safer States (2010) Dispersants debate in Gulf spill highlights need for TSCA reform, 17 June. Available at http://www.saferstates.com/2010/06/oil.html.

Schor, E. (2010) EPA chief calls for more authority over dispersants, *New York Times*, 15 July. Available at http://www.nytimes.com/gwire/2010/07/15/15greenwire-epa-chief-calls-for-more-authority-over-disper-72278.html.

Stier, C. (2010) Gulf oil spill poses unique health challenges, *Coastal Care*, 9 July. Available at http://coastalcare.org/2010/07/gulf-oil-spill-poses-unique-health-challenges/.

Wechsler, P., & Gross N. (2010) Health data gaps, BP suspicions worry US panelists, *Business Week*, 22 June. Available at http://www.businessweek.com/news/2010-06-22/health-data-gaps-bp-suspicions-worry-u-s-panelists.html.

Weinberg, D. (2010) Louisiana's fragile indigenous Gulf tribes face uncertain future, *Voice of America*, 17 July. Available at http://www.eurasiareview.com/201007175197/louisianas-fragile-indigenous-gulf-tribes-face-uncertain-future.html.

A Gap Hazard Analysis: Initiating Policy Development with Mountainous Communities

G. STRICKERT*, S. SMARASINGHE*, C. DOSCHER* & T. DAVIES**

*Department of Environmental Management, Lincoln University, Canterbury, New Zealand; **Department of Geology, University of Canterbury, New Zealand

ABSTRACT *We present a conceptual framework and methodology for policy development research in complex, uncertain, and data-poor environments with the objective of sustainable hazard mitigation. Creating effective policies for improving resilience to natural catastrophes in general and developing policy initiatives to help manage earthquakes in particular requires a conceptual framework whose formulation should engage lay-experts and experts as essential members of the research process. This paper outlines a framework and applied case study as a starting point for identifying policies that increase resilience to earthquakes or other low probability high consequences events. The methodology integrates geomorphic assessments, qualitative interviews, and fuzzy cognitive maps with three stakeholder groups for simulating policies scenarios. The results indicate that an integrated gap hazard analysis embedded in a policy framework for increasing resilience to earthquakes in New Zealand ski areas does identify 'gaps' in stakeholders' hazard perceptions and awareness. Furthermore, the framework is a useful first step toward policy development for hazards beyond current horizons.*

Introduction

Development of effective policy for sustainable hazard mitigation is a complex process. The process has the characteristics of a complex adaptive system including: adaptive memory and feedback (i.e. in both a social and physical sense), the non-linear relationships (i.e. inputs and outputs are not proportional) the process can adapt itself and its environment, and the system is sensitive to initial conditions (Ostrom, 1999; Corbacioglu & Kapucu, 2006). In addition, the dynamically uncertain nature of hazards can limit our understanding about them which challenges our ability to forecast their occurrence. This requires that people at risk need to be prepared to respond and recover effectively. As policy development for hazard mitigation involves communities that have experienced (or will experience) these events, it is prudent that an adaptive and participatory approach be taken to address this problem. This study focuses on demonstrating the effectiveness of the approach for indentifying key policy considerations for hazard mitigation in the context of ski-fields (alpine ski areas) prone to earthquakes.

The study of natural disasters and hazards 'transcend the exact methods associated with traditional "hard" sciences' stressing that to deal with complexity and uncertainty, "'softer", and more flexible methods and tools are required' (Weichselgartner, 2001, p. 2). This calls for a transdisciplinary mixed-method approach for natural hazard and disaster research. An effective way to apply such an approach is to work towards practical strategies (policies, guidelines, etc.) that are developed through stakeholder participation (Mileti, 1999) and driven within existing communities of practice (Wegner *et al.*, 2002). In achieving this, Mileti (1999) also highlighted a new paradigm in natural hazard management, namely managing natural hazards from a sustainable development agenda; or 'sustainable hazards mitigation', a term he (with the help of many others) coined.

Sustainable hazard mitigation is achieved when 'public participation is integrated into disaster management planning and community planning' (Pearce, 2002). Yet Pearce (2002, p. 218) highlights Dorcey and McDaniels (1999) who 'document the plethora of multi-stakeholder consensus processes that proliferated in the 1990s, . . . but were criticized as being too costly, slow, overemphasizing public interests, and usurped the role of elected officials'. Dorcey and McDaniels (1999) also argued that consensus-based approaches may be desirable, but they can heighten expectations of empowerment for actual decisions when their stakeholders' actual role is advisory. 'Stakeholder participation needs to be underpinned by a philosophy that emphasizes empowerment, equity, trust and learning. Where relevant, participation should be considered as early as possible and throughout the process, representing relevant stakeholders systematically' (Reed, 2008, p. 2417).

Many theoretical approaches have been developed that attempt to optimize stakeholder involvement to improve policy development. For example, consider the lack of a single definition for participatory disaster risk assessments (PDRA) which attempts to engage stakeholders in the assessment of hazards, vulnerability, coping capacity, and risk (Pelling, 2007). The lack of a definition can reflect the diversity of interests from agencies who are interested in PDRA. Though, as Pelling (2007, p. 375) states 'a lack of common understanding also opens this field of work to misplaced or exaggerated claims of participation, inclusiveness or, perhaps most difficult of all, empowerment'.

Relevant to this study are also the risk management approach, the empowerment approach, and the resilience approach. The strengths of the risk management process is that it has become more standardized (i.e. AS/NZ 4360, HB 167: 2006, and ISO 31000), and simpler to follow (see Gibson, 2007). Despite standardization and simplicity it has a myriad of possible instruments, tools, and frameworks to improve participatory policy development. However, participatory risk management can fall victim to the 'tool-kit' approach failing to emphasis participation (Reed, 2008). The overall strengths of the empowerment approach are due to the comprehensiveness over earlier theories gained from the more recent alignment of the two perspectives: the social-structural and the psychological (Spreitzer, 2007). It has also been proposed that empowerment can be integrated with other areas of research such as health, leadership and resilience (Spreitzer, 2007), and there are calls for longitudinal research on the integrated perspective of this approach (Conger & Kanungo, 1988; Thomas & Velthouse, 1990; Spreitzer, 2007). A criticism of the empowerment research is that its measurement of empowerment relies on extractive methods data of collection. The resilience approach has received traction in recent years because it resonates with uncertainty and complexity which favours adaptive capacity (Walker *et al.*, 2002).

However, it may be difficult for many organizations to shift to adaptive management and resilience ideologies as they do not necessarily provide clear paths toward future policies.

Context: Natural Hazards and New Zealand's Alpine Ski Areas

Alpine Ski Areas are responsible for managing any 'potential cause or source of harm' (Peter & Priestley, 2004, p. 4). The current culture of safety in Alpine Ski Areas functions very well with regards to reduction and readiness of, response to, and recovery from higher probability low consequences events (e.g. physical injuries, severe weather, and avalanches). However, according to D. Masey (personal communication, August 2008), the management of low probability high consequence events (e.g. earthquakes) is not well-established in the ski industry, though the management of lahars is an exception. The current level of management of the risk of lahar at Ruapehu was initiated due to the 1969 and 1975 eruptions, which were the main examples that had been studied (Otway *et al.*, 1995). The former eruptions occurred with little warning (i.e. only about 10 minutes of seismic activity); hence, their unpredictable nature requires that they are handled similar to earthquakes where the risks can be reduced by pre-emptive coping strategies (Otway *et al.*, 1995). Ongoing research on the lahar risk at Mt. Ruapehu ski fields has helped develop and improve generalized models for hazard warning systems and improving staff and public hazard education and emergency response planning (Galley *et al.*, 2004).

The sophisticated studies on the lahar risk at Ruapehu were initiated due to previous lahars and eruptions on Ruapehu. New Zealand ski fields in general face challenges for mitigating the risks associated with natural hazards in particular managing transient populations during emergency response (Leonard *et al.*, 2008). New Zealand ski fields have yet to directly experience a major earthquake (e.g. > Magnitude 7), though at some point in the future it is certain to happen (Yetton *et al.* 1998; CDEM, 2002a, 2002b). A few earthquakes have impacted the ski areas since the first ski area was established in 1913. For example, in 1994, a driver of a snow grooming machine was killed in an aftershock which triggered an avalanche (Owens *et al.*, 2002). Several other earthquakes have affected the ski fields in New Zealand, but fortunately the impact on humans and infrastructure in the ski areas has been limited. The magnitude 7.8 Fiordland earthquake of 15 July 2009 can be considered a near miss, as it ruptured in such a manner as to cause a slow rolling motion instead of intense shaking which translated to little damage (GeoNet, 2009). There are currently 26 ski areas in New Zealand; the approach presented here is a first step at identifying opportunities for managing the earthquake risks and their consequences prior to experiencing an earthquake.

In an effort to improve the mitigation of natural hazards, New Zealand has put sustainable hazard mitigation within their Civil Defense and Emergency Management (CDEM) Act 2002 which replaced the Civil Defense Act 1983. The Preliminary Provision of the Act states that one of the key purposes is to '. . . improve and promote the sustainable management of hazards . . . in a way that contributes to the social, economic, cultural, and environmental well-being and safety of the public and also to the protection of property; . . .' (CDEM, 2002a, p. 7). The CDEM Act (2002b) stipulated that cross-boundary and regional CDEM groups had to be established and to consult with local communities in the development of CDEM plans.

New Zealand's Resource Management Act (RMA) 1991 also supports sustainable hazard mitigation; the purpose of the Act is the sustainable management of natural resources.

Saunders *et al.* (2007) outline a framework to help strengthen the linkages between the CDEM Group Plans and RMA plans emphasizing Regional Policy Statements and the need for incorporating reduction measures and actions in the CDEM Group Plans. Kingsbury (1996) also highlighted the importance of Regional Policy Statements that can enable progress of the Earthquake and Geological Mitigation Strategy for Wellington City, but asserted that effective hazard mitigation requires 'translating good science into a useable form and transferring it to non-technical users' (Kingsbury, 1996, p. 28).

The Australia and New Zealand Standard (AS/NZ 4360:2004) is also highly relevant to this study. This risk management standard is outlined in the Australian Capital Territory Insurance Authority *Guide to Risk Management* (2004). Critical in the guide is the need for Stakeholder Consultation/Communication within all steps of the risk management process. Moreover, the AS/NZ 4360 standard became the foundation for ISO/FDIS 31000: Risk Management—Principles and Guidelines.

Overview and Rationale of the Triangulated Approach

In social systems, the existence of institutions and networks that learn and store knowledge and experience, creates flexibility in problem solving, and balance of power among interest groups play an important role in adaptive capacity (Scheffer *et al.*, 2000; Berkes *et al.* 2003; in Resilience Alliance, 2007).

Our study demonstrates *methodological triangulation* (Denzin, 1978) using a *development* mixed method approach (Green *et al.* 1989) that identifies, analyses, and evaluates risks, and identifies possible treatments for the risks within an integrated participatory framework. Methodological triangulation is the use of multiple methods to study a research problem (Tashakkori & Teddlie, 1998). A development mixed method approach applies different methods sequentially such that the first method informs the second and so on (Greene *et al.*, 1989). Triangulated methods can enable the weakness of one method to be offset by strengths of the others (Jick, 1979). Basically, the gap hazard analysis involves an integrated set of tools and conceptual framework that helps to identify policy opportunities for mitigating natural hazards that get little consideration.

One of the essential theoretical underpinnings of this problem-focused mixed-methodology is an attempt to apply sustainable hazard mitigation (SHM) on a practical level from the outset of the research agenda.

The gap hazard approach fosters stakeholder participation to identify policy opportunities by integrating three sequential methodologies: (1) assessment of the physical context of hazards using geomorphic assessment; (2) qualitative interviews with the aim of probing into experiences and perceptions; and (3) developing fuzzy cognitive maps (FCM) that capture the perceptions of the participating stakeholders (Strickert *et al.*, 2009). Once the data is gathered and organized into social cognitive maps, it can be modelled using an auto-associative neural networks (AANN) framework for simulating policy scenarios to draw inferences from FCM (Kosko, 1987; Kosko, 1992; Reimann, 1998; Ozesmi, 1999; Ozesmi & Ozesmi, 2003). In this method, the outcome of an FCM can be determined with "matrix algebra where a vector of initial states of variables (I^n) is multiplied with the adjacency matrix of *A* the social cognitive map" (Ozesmi & Ozesmi, 2003, p. 523). Next a threshold function, in this study it was a logistic function $1/(1 + e^{-1*x})$, is used to transform the results to the interval zero or one. Then the new vector-state is again multiplied by the adjacency matrix and transform by the logistic function iteratively until the system stabilizes, falls into an oscillation, or goes into chaos (Disckerson & Kosko, 1994; Ozesmi & Ozesmi, 2003).

There are three main reasons for the integrated approach. First, it was hypothesized that people in ski areas do not overtly consider the risk of earthquakes and their potential consequences. Hence, the approach had to examine what the possible consequences could be, which were gathered using geomorphic assessments. For example, an earthquake in a mountainous area can cause any isolated or combination of Sturzstroms, landslides, snow avalanches, and rock fall (Sharpe, 1960; Varnes, 1978; Saunders & Glassey, 2007). The assessments lie on a continuum between *superficial*, looking for generic and emergent consequences across similar landscapes, and *intense*, specialized investigations of a particular hazard limited to one area.

With the data from geomorphic assessments, the researchers had a choice whether to continue with quantitative surveys, or to use in-depth qualitative interviews as a next step. Qualitative interviewing was chosen as it gave the researcher the opportunity to actively engage the participants and when needed probe deeper into specific areas of interest that emerged. The interviews cultivate research that 'attempts to understand the world from the subjects' point of view, to unfold the meaning of peoples' experiences, to uncover their lived world prior to scientific explanations' (Kvale, 1996). The quotations from the interviews reveal the way participants have internalized the world including their thoughts about what is happening in their world, their experiences, and basic perceptions (Patton, 1987). Qualitative interviews also facilitate the development of fuzzy cognitive mapping, particularly, if the research problem is one that participants have never considered (Strickert *et al.*, 2009).

Qualitative interviews do not afford the capacity of simulation which can enhance the policy development process. Therefore, a complimentary data gathering technique called Fuzzy Cognitive Mapping (FCM) (Kosko, 1987; Ozesmi, 1999) becomes useful. 'FCM is a soft-system modeling and mapping approach combining aspects of qualitative with the advantages of quantitative methods. It has a strong participative component fostering social learning between the participants' (Bachofer, 2009). Though techniques with similar capacity exist in participatory policy development in hazard and risk management; for example, structural equation modelling (McIvor & Paton, 2007), multi-attribute decision theory, multi-attribute value theory (Bertsch *et al.*, 2007), and Bayesian belief networks (Trucco *et al.*, 2008), 'Cognitive maps can serve as a basis for discussion when policies and management options are formulated' (Ozesmi, 1999, p. 144). An FCM is a qualitative model of how a system operates and is especially useful for modelling complex relationships among variables. Particularly, FCM is useful in four types of problems where gaining insights or predicting system behaviour is not obvious or easy:

> (1) prediction which involves human behaviour and how human actions can unknowingly affect ecosystems . . . (2) where detailed scientific data are lacking but local knowledge of people adapted to an ecosystem is available . . . (3) 'wicked' environmental problems . . . are complex, involve many parties, and have no easy solutions or right answers . . . (4) issues in ecosystem management where public involvement is desired or even mandated by law. (Ozesmi & Ozesmi, 2004, pp. 43–44)

The multi-step approach is a synthesis of relevant useful techniques from many disparate disciplines on cognitive mapping for analyzing how people perceive an ecosystem, and for comparing and contrasting the perceptions of different people or groups of stakeholders (Ozesmi & Ozesmi, 2003). The FCM can be used to develop a participatory management plan based on stakeholder input where the stakeholders are able to take ownership of the plan and hence can work towards its goals.

In the multi-step approach, FCM are condensed and aggregated prior to simulations Ozesmi and Ozesmi (2003). Typically, participants generate a large number of variables (key considerations) some of which are highly related and therefore, are amenable to condensation into high level concepts and considerations (Nakamura *et al.* 1982). This simplifies the subsequent model development and enhances the understanding of the outcomes. The condensation is a subjectively qualitative process, categorizing the conceptually similar key considerations in respondents' FCM into high level categories (Ozesmi & Ozesmi, 2004**;** Strickert *et al.*, 2009). Furthermore, individual's FCM can be aggregated into Social Group Fuzzy Cognitive Maps (SGFCM) of stakeholder or participant groups (Laszlo *et al.* 1996). Once FCM are processed this way, realistic policy options can be simulated where the SGFCM provides the structure for recurrent evaluation (i.e. an auto-associative artificial neural networks (AANN)) (Kosko, 1992; Reimann, 1998; Ozesmi, 1999; Ozesmi & Ozesmi, 2003; Strickert *et al.*, 2009). These networks do not learn in the same way as traditional neural network architectures (e.g. Hebbian or Backpropogation), rather the researcher becomes the interpretive tool of the network in a 'steady state'; or more aptly a 'desirable configuration'. Each neuron in an ANN represents a key consideration and the connection strengths between the neurons in the FCM are the same as those provided by the participants for the strength of relationship between the corresponding key considerations.

Each neuron in the network can be depicted as shown in Figure 1 to demonstrate its operation.

A neuron as shown in Figure 1 represents a variable (i.e. key consideration). Each neuron receives inputs (I_n) from other neuron and itself. Inputs are the level of importance given to each of the key considerations, in the range of [0 to 1]; 0 meaning no importance whatsoever is given to that consideration and 1 means the highest importance is given to it. Initially, these inputs reflect the status quo and can be ascertained from participants. Each input is weighted by the strength of the connection (w_n) between the neuron sending the input and the neuron receiving it. Thus all inputs received by a neuron are weighted by the corresponding connection strength, and the neuron aggregates the

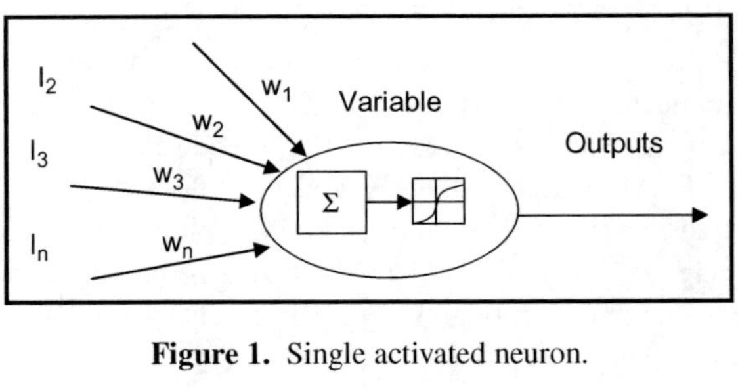

Figure 1. Single activated neuron.

weighted input by summation (Σ). The neuron can produce this as its output, but a better approach is to standardize or normalize it so that the output always stays within the bounds of 0 and 1. This is achieved by processing the net sum by a logistic function. This output is then sent to all other neuron and itself as input and the process repeats until there is no change to the outputs, meaning that a steady state or equilibrium has been reached (usually within 30 simulation steps or loops). The steady state reflects the effect of initial inputs (level of importance given to key considerations) on the net outcome of the whole system.

Once the steady state solution to status quo is ascertained, the neural networks can be used to simulate the effect of changing the status quo to a desired policy setting with respect to the level of importance given to the key considerations (inputs) in the network. For example, by setting all inputs to 1 means a policy where all key considerations are given high priority. The change in the outcome with respect to that of the status quo gives an indication of the effect of the change. Alternatively, by changing each input individually to a desired level and leaving others at the level of status quo, the importance of that input to the whole system (sensitivity) can be ascertained. By changing all inputs to various levels, a variety of policy options and their effects can be simulated.

For instance, consider a conceptual policy scenario example: the parking lots in a mountain tourism area are a safe zone (i.e. *Physical Preparation*), and a major landslide (i.e. *Adverse Physical Effects*) damages the safe zone when physical effects are set at high level (i.e. Input = 1). During each step of the simulation *Adverse Physical Effects* is set at 1. Then the simulation output of this policy scenario is compared with the original steady states (status quo). The participant(s) may have indicated that when *Adverse Physical Effects* are high (i.e. major landslide), this has a significant influence on *Physical Preparation* (i.e. damage to the safe zone).

Mixed-Method Multi-Site Case Study

The case study started by conducting geomorphic assessments at six ski areas during the summer season to determine the possible consequences of earthquakes with respect to topography, geology, and infrastructure. The assessments helped develop targeted questions for qualitative interviews and inferences for scenario simulations. Subsequently, qualitative interviews were conducted which consisted of semi-structured qualitative interviews and fuzzy cognitive maps (FCM). This study adapted Ozesmi and Ozesmi (2003) multi-step fuzzy cognitive mapping approach. Participants developed an FCM of the critical variables (important considerations) about preparation for, response to, and recovery from an earthquake that has a direct effect on the ski areas. Specifically, once they identified critical variables, their attention was shifted to a series of questions about their perceptions and experiences with catastrophic natural events (i.e. natural disasters, severe weather, mass movements, and earthquakes). Once the questions were complete, the interviewer returned to the FCM asking if they wanted to add any more variables. Once satisfied with their selected variables, they were asked to assign critical connections with strength values in the range of [-1, 1] (-1 indicates a strong negative effect and +1, a strong positive effect). Once satisfied with their FCM, participants were asked to sign it, which completed the interview (Figure 2).

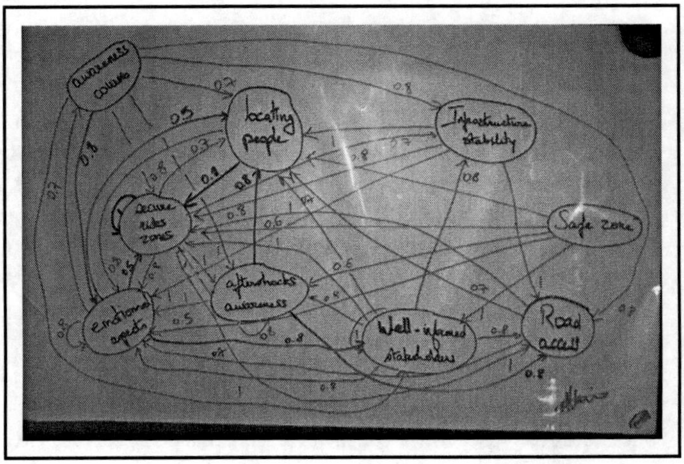

Figure 2. An example FCM from the case study demonstrating the key considerations for preparation, response and recovery and their relationships in relation to an earthquake that affects a ski area.

Sampling Rationale

The ideal scenario for sampling would have been stratified random sampling of stakeholders at all 26 of New Zealand ski areas, however, the scope of the study was limited in funding and resources as it was a PhD study. Within the study there was also convenience and snowball sampling. Convenience sampling occurred when interviews took place both on mountain and off mountain in various accommodation centres après ski. Participants in these instances were approached using a research information script which was approved by the Lincoln University Human Ethics Committee in April 2007. This was the most effective manner to gather lay-expert participants. Snowball sampling occurred often after interviews with managers where they suggested participants within theirs or other organizations who would add value to the study. This form of snowball sampling was highly valuable as it enabled the researcher to interview key staff members responsible for safety and emergency services at various ski areas.

The purpose of the sample was to demonstrate an application of FCM within a novel context and framework. In total there were 52 participants originating from 12 different countries. Forty-two interviews took place in New Zealand while 10 took place in Canada. The Alpine Snow Riders (ASR) group consisted of 13 participants (both skiers and snowboarders) with two experts and 11 lay-experts across the full range of experience levels from 6 distinct ski areas. The Ski Industry Workers (SIW) group consisted of 19 participants' with 12 experts and seven lay-experts of medium to high levels of experience including: ski instructors, safety services, ski patrollers, services staff, technicians, avalanche controllers, and operations staff from 12 distinct ski areas. The Ski Area Managers (SAM) group consisted of 15 expert participants including: mountain managers, operators, and owners from 14 distinct ski areas. A more diverse sample would have enabled the results to be generalized beyond the context of alpine ski areas. Hence, in future research more diverse samples should be sought. Given the purpose of the sample, the qualitative sample was sufficiently stratified, random, and displayed reasonable heterogeneity.

Results: Geomorphic Assessments, Qualitative Interviews, and Fuzzy Cognitive Maps

Geomorphic Assessments

The main geomorphic issues as related to co-seismic events from large to small were: Sturztroms, landslide, snow and rock avalanche, rock fall and infrastructure collapse or damage including: reservoir failure, road section collapse, building collapse, and loss of critical lifelines. Emergent findings from the geomorphic assessments were the critical locations of large water reservoirs and lakes used for snowmaking. Some of these water bodies could be subject to damage and possible collapse in high consequence events. The outflow paths from the reservoirs run through critical infrastructure with a reasonable possibility of casualties, both in lives lost and property. This potential scenario is aptly termed 'disaster(s) by design' (Mileti, 1999) (Figure 3). The possible mechanisms for reservoir collapse are: displacement of water via seismically induced rock or snow inundation into water bodies, reservoir wall failure, and the potential for sieching during seismic events. Seismic seiche 'a term first used by Kvale (1953) in discussing oscillation of lake levels in Norway and England caused by the Assam earthquake of August 15, 1950' (McGarr & Vorhis, 1968). Serway and Jewett (2006) explain that 'During the Northridge earthquake of 1994, swimming pools throughout Southern California overflowed as a result of seiches set up by the shaking of the ground' (p. 451). Interestingly, these reservoirs might help in monitoring the earthquakes (McGarr & Vorhis, 1968) and be a helpful tool for studying topographic amplification.

Topographic amplification occurs near peaks or ridges. During an earthquake, maximum ground acceleration (i.e. 'shaking') is most pronounced at peaks and ridges (Buech, 2008), similar to how earthquakes affect buildings (the top of the buildings move more than the bottom). In mountains, the process is more complex, intense, and compounding than for engineered structures. Development in mountainous areas can place high numbers of people at an unnecessary and involuntary risk.

Results from Qualitative Interviews

Most participants had never considered the case of an earthquake while they were in a ski area. Subjects were asked various open-ended questions about experiences, and

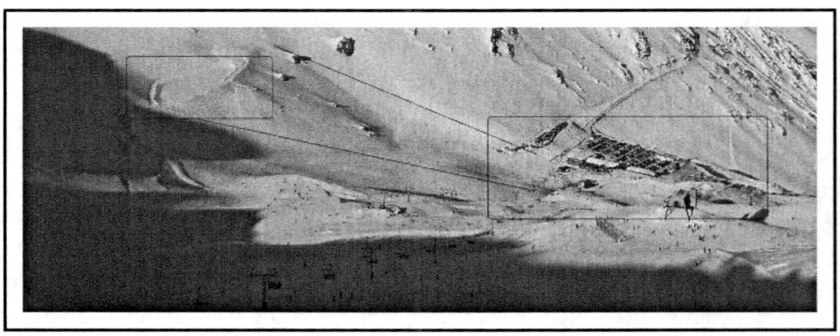

Figure 3. Reservoir located above critical infrastructure.

perceptions related to natural disasters in general, and more specific probing questions about their experiences with severe weather, avalanches, and finally, the targeted event; an earthquake.

With respect to earthquakes and their consequences, participants consistently responded that these events were not in the participants' horizons. Here, two managers of different ski areas explain:

> It hasn't been something that really has caused an issue in any ski area that I know in New Zealand that we have a record of, you know. So it hasn't been one that has been at the forefront of my mind for sure. (Commercial Ski Field Manager # 6)

> Um, well we have um, when people arrive we ask them to be equipped for any event. Um, it's a difficult one, cause we really, it's sort of, it's one of those things we don't really plan for. Because it's um, hum? (Club Ski Field Manager #1)

Ski Area workers, particularly those concerned with snow safety (i.e. avalanche control, and first aid) also highlighted how *earthquakes* are not key considerations in their day to day work:

> Once or twice I found myself kind of well I wonder if, are we . . . are we missing *that* hazard out of our overall planning. And um . . . how would it affect me? (Ski Industry Worker #1, Safety Services)

> I don't think anyone's really put much thought to it if a big earthquake happened. (Ski Industry Worker #4, Ski Patrol Director)

While ski field users suggest similar notions:

> No, not an earthquake, I have never considered that. (Alpine Snow Rider #8)

> I am not aware of any ski area in the country or, in fact, the world, in fact, that I've been to, where earthquake has ever been mentioned in any material that I've come across. (Alpine Snow Rider, #1)

Despite the apparent lack of considerations of earthquakes and their effects on ski areas, participants did provide constructive insights into their perceptions of key considerations about preparation for, response to, and recovery from earthquakes that affect a ski area (Strickert *et al.*, 2009). The majority of the participants perceived that there would be adverse physical effects from an earthquake on the ski area including: avalanches, rock fall, and damage to buildings, and aftershocks that could pose major challenges to the response efforts. When imagination of the physical effects manifested in the minds of participants, many were able to detail flow-on effects from the initial damage. Most outlined that existing multi-hazard emergency procedures would be able to effectively manage an earthquake; though some participants did note that the ability to respond was conditional on the severity of the earthquake. But, the fickle nature of earthquakes, that is, the inability to precisely locate where, when and at what magnitude they are likely to happen, presented a dilemma to the participants, making it difficult to detail whether they would be able to manage the effects of an earthquake.

Results from Fuzzy Cognitive Maps

Condensation occurred prior to aggregating FCM into Social Group Fuzzy Cognitive Maps (SGFCM) of different participant groups. 'The condensation enables aggregation of 'similar' variables into upper level categories' (Strickert *et al.*, 2009). In this study, almost three hundred variables were condensed into 10 categories of key considerations (Table 1). Once FCM were condensed into categories the individual maps were aggregated into SGFCM. Then the structure of the SGFCM was explored 'by finding transmitter variables (forcing functions, givens, heads), receiver variables (utility variables, ends, tails), ordinary variables and immediate domain (or centrality of variables)' (Bougon *et al.*, 1997; Eden *et al.*, 1992; Harary *et al.*, 1965, as cited in Ozesmi, 1999, p. 151). These variables are defined by their outdegree (total strength of connections going out of a node) and indegree (total strength of connection coming into the node). The immediate domain (a.k.a. total degree or centrality) is the summation of outdegree and indegree, and is an indication of the 'importance' of a single variable in a cognitive map (Strickert *et al.*, 2009; Ozesmi, 1999). Transmitter, receiver and ordinary variables were calculated using the ratio of outdegree to indegree. Transmitter variables' ratio was set at greater than 1.055, receiver variables' ratio at less than 0.95, whereas, that of ordinary variables was set between 0.96–1.05. The resulting classification of variables are shown in Table 2.

Overall, these results only indicate *perceived* trends in participants' SGFCM; that is, they are a product of condensation, aggregation and the threshold ratios set between outdegree and indegree. In this configuration, it appears that the managers' maps contained more of the upper level considerations as transmitter variables, followed by receiver variables, and then ordinary variables. Arguably, this means that as a collective social group they may sense some degree of control over the ski area in response to an earthquake. The results from the ski industry workers (SIW) contained the same number of transmitter variables as ordinary variables followed by receiver variables. This same number of ordinary variables

Table 1. Upper level categories (condensed variables) and examples of critical considerations in them

Upper Level Category	Examples of Critical Variables in Participants FCMs
Lifelines	food and water, road access, communication, and energy
Negative (Adverse) Physical Effects	aftershocks, avalanches, rock fall, and building collapse
Physical Preparation	safe zones, rescue equipment
Physical Recovery	certified engineers' assessments and rebuilding infrastructure
Physical Response	chair-lift evacuation, and immediately assessing infrastructure
Social Recovery	reflecting, debriefing, and updating emergency procedures
Social Preparation Action (SPA)	hazards awareness courses, training, and buddy system
Social Preparation Resources (SPR)	communication tools, customer involvement, and emergency responders
Social Response Action (SRA)	effective communication, medical attention, and evacuation
Social Response Resources (SRR)	emergency contacts, personnel, and procedures

Source: After Strickert *et al.* (2009).

Table 2. Classification of transmitter (T), receiver (R), and ordinary (O) variables on the FCM of managers, ski industry workers (SIW) and Alpine snow riders (ASR)

Variables	Managers	SIW	ASR
Lifelines	R	R	O
Physical Effects	T	R	R
Physical Preparation	T	T	R
Physical Recovery	O	T	T
Physical Response	T	T	O
Social Recovery	R	O	T
SocPrep Action	O	T	O
SocPrep Resources	R	O	R
SocResp Action	T	O	R
SocResp Resources	T	O	O

Source: After Strickert *et al.* (2009).

and transmitter variables may indicate a nuanced horizon that requires adaptation, rather than control in the face of an earthquake. The alpine snow riders (ASR) results are different, in that they displayed the same number of receiver variables as ordinary variables, followed by a fewer number transmitter variables. It can be inferred that the ASRs perceive themselves as affected by their environment, instead of in control of the environment, which is quite different to the managers.

Simulation Results of Key Policy Consideration Scenarios

The simulation results for the maps of Total Social Group FCM (TSGFCM) and each of the three participant groups are shown in the four charts in Appendix A, B, C, and D. Each chart shows the influence of 10 simulation scenarios of each policy consideration (i.e. each key consideration in the FCM) on all of the other upper level policy considerations. This is done by setting one policy consideration at a high level for all the simulation steps until the steady state is reached. This process is repeated for the other desired policy considerations. Accordingly, the bar charts in the Appendices are colour coded showing the influence of setting one policy scenario at a high level on the other policy considerations (Note that the colour coded Appendices are only available in the electronic version). The legend on the right of the charts represents policy considerations that were given high priority level of 1. The effects of the individual policy consideration on the other key considerations are represented by the degree of influence depicted on the vertical axis as the percent change (upward movement) from the steady state level for the same considerations with respect to status quo. For example, the first rows of blue bars indicate the influence of setting *Lifelines* at a high level on the other considerations. At this first stage of simulation, all of the variables (key considerations) are treated as independent variables.

TSGFCM Key Policy Considerations (See Appendix A)

The TSGFCM simulations of policy considerations supported the physical nature of the overall perceptions of the participants. All of the simulation options had some influence on *adverse physical effects*. That is, the TSGFCM indicates that all of the upper level policy considerations may reduce the adverse physical effects such as earthquake induced mass movements and damage to infrastructure. Notably, *Physical Recovery*, for example rebuilding infrastructure, had the highest influence on *Adverse Physical Effects*. This might mean that collectively they hold a 'long-term view' or cyclical horizon where *Physical Recovery* would allow them to decrease the *Adverse Physical Effects* during the next CNE by relocating infrastructure and improving seismic constraints during new construction. Also, when *Adverse Physical Effects* were high, this affected *Physical Response*, *Social Recovery,* and *Social Preparation Resources*. Hence, the belief held by many participants that the scale and or intensity of the event impact the preparation, response, and recovery.

Ski Area Managers (SAM) Key Policy Considerations (See Appendix B)

The SAM's most noticeable policy scenario was that setting *Social Prepration Resources* (i.e. Hazard Awareness Courses) at a high level had a high influence on *Social Recovery* (i.e. debriefing and counseling), *Physical Recovery* (i.e. rebuilding infrastructure), and *Lifelines* (i.e. energy, transport, communication). The perceptions of the managers also indicate that if *Social Recovery* is at a high level, then *Social Preparation Resources* would be significantly influenced, demonstrating that the managers perceive a long-term view, whereby, if they recover socially from a CNE, mangers may allocate more resources to preparing for future events. When *Physical Recovery* was set at a high level this also had an influence on *Social Preparation Resources* and *Social Recovery*. In fact, in of all the simulations that were run, the most noticeable influences from the managers were on recovery considerations. This might translate to managers perceiving that, with respect to earthquakes, recovery is most critical to ensure they can maintain their livelihood through operating the ski area.

The Ski Industry Workers (SIW) Key Policy Considerations (See Appendix C)

The scenario simulations for SIW highlight that *Physical Recovery* had a significant impact on *Physical Effects and Lifelines,* again possibly showing the 'long-term view' horizon with feedback and the cyclical nature of their perceptions toward CNE. The SIW horizon was similar to the TSGFCM in that most policy scenarios had an impact on *Adverse Physical Effects*. They also perceived that if *Adverse Physical Effects* were high, *Physical Response* (i.e. lift evacuations), *Social Response Resources* (i.e. emergency management systems), *Social Preparation Resources,* and *Lifelines* (i.e. communication, energy, water and transport) would be impacted. Furthermore, this simulation implies that high *Adverse Physical Effects* would have an impact on the social considerations, thus strengthening remarks regarding conditional response and recovery being dependent on scale and intensity. When *Physical Response* (i.e. Immediately Assessing Infrastructure) was high, the SIW perceived improvements in *Physical Recovery* (i.e. rebuilding infrastructure) and *Social Response Resources* (i.e. emergency personnel for future events), with a marked infuence on curtailing *Negative Physical Effects* such as mass-movements.

Alpine Snow Riders (ASR) Key Policy Considerations (See Appendix D)

The ASR simulations also noted the high influence of *Physical Recovery* (i.e. rebuilding infrastructure) on *Adverse Physical Effects* (i.e. consequences of earthquakes), again possibly showing the 'long-term view' horizon with feedback and the cyclical nature of their perceptions toward CNE. Another similarity was that most strategies had an impact on *Adverse Physical Effects*, meaning that they share similar horizons with the TSGFCM. The ASR also perceived that if *Adverse Physical Effects* were high, *Physical Recovery and Response* and *Social Recovery* would be impacted. The high influence on recovery is similar to the managers, but may be more indicative of maintaining their recreation and leisure time. The relatively high influence of this simulation may indicate a perceived need to recover. Also, this simulation demonstrated that high *Adverse Physical Effects* would have some influence on *Social Preparation Resources* and possibly reduce the efficacy of *Physical Preparation.* Another difference was that when *Physical Response* was high, the SIW perceived impacts on *Adverse Physical Effects*, perhaps a perception of control that was not noticeble prior to simulation. ASR also perceived *Social Response Resources* had infuence on curtailing *Negative Physical Effects*.

Discussion

The approach taken in this case study was an extractive participatory approach. It tried to achieve two difficult and perhaps mutually exclusive goals: (1) indentify the risks and hazards associated with earthquakes in ski areas; and (2) apply an integrated framework including an adapted and novel use of FCM. Our approach is not intended to be a haphazard replacement of the risk management approach or the empowerment approach, or the resilience approach; rather to compliment them by engaging stakeholders using FCM as a precursor to more emancipatory, conscientious, and longitudinal approaches. Weakness of other approaches could be overcome by using the FCM tool presented in this framework or improvements thereof.

The strength of the approach is that it provides a rigorous conceptual framework, augmented with computational algorithms and powerful methodologies. The latter extends the adaptive management/resilience paradigm by identifying policy opportunities where trial and error can occur in simulation instead of in the real world, where adverse consequences of policies can be severe. The challenge in the case of earthquakes and ski fields is that the managers, the workers, and the users are not overtly aware of the earthquake hazards. Hence, the mixed method proposed herein is not intended to replace other forms of engagement, rather it is an initial form of consultation/communications that can enhance and accelerate the social learning required for later participatory policy development, particularly when more advance neural networks are employed.

The scenario simulations conducted here are not in the traditional form of neural network simulation as no learning is taking place within the network. However, these initial simulations can be used to identify targets for more advanced learning-based simulations within the neural networks paradigm. For example, for a 'desirable configuration' (targeted) policy option, the network can be evolved from its status quo to the desired response policy by adjusting the connection weights using a learning network. These new weights can then inform ongoing engagement with the stakeholders to understand their perceptions better, thus enabling a more synergistic transformation of the expert and lay-expert perceptual models to better inform policy. This is beyond the scope of this paper, but will be the focus of our forthcoming paper (Samarasinghe & Strickert, 2011, in press).

With respect to the policies mentioned above, we believe that the framework presented can enhance the policy-making process. Fundamental to advancing this end, the process of gathering, analyzing, and interpreting FCM needs to be standardized and made even more transparent. This can be achieved through prolonged engagement with participants while undertaking several iterations of the mapping process within existing conceptual frameworks. Simulations can be run with wide varieties of participant groups beyond the three that were targeted in this study. For example, social groups can be created from demographics, expert and lay-expert groups, or creating SGFCM from people with similar responses to qualitative questions. In the New Zealand context, such a standardized framework would enable more synergies to be identified between the RMA and the CDEM. Moreover, FCM may be used as a means to develop Regional Policy Statements that are stakeholder driven thus overcoming issues of legitimacy, and can focus more on reducing the impact of natural hazards.

Conclusions

We presented a conceptual framework and methodology for policy development research in complex, uncertain, and data-poor environments with the emphasis on sustainable hazard mitigation. A mixed-methodology was applied for indentifying gaps in hazard perception and awareness and developing policies to manage earthquakes in mountain landscapes. Ski areas provided context and communities of practices for triangulated mixed-methodologies. Experts and lay-expert participants provided rich data about the experiences and perceptions towards hazards in mountainous environments. All three groups perceived that developing strategies in preparation for, response to, and recovery from earthquakes could reduce their adverse consequences. Furthermore, it was perceived that experiencing an earthquake would lead to greater resilience in the future. The results indicate that the gap hazard analysis, embedded within a policy framework for improving resilience to earthquakes in New Zealand ski areas, does work well to identify gaps in stakeholders' hazard perceptions and awareness. Furthermore, the framework is also a good starting point for indentifying opportunities to improve policy for managing low probability high consequence events. Refining the mix of methods used in this case study can advance sustainable hazard mitigation and aid participatory approaches in vital areas of research.

Acknowledgements

We would like to thank the following: our participants in the study for giving us your time and insight about your lived experiences and living perceptions; the ski areas who gave us access to your sites for all stages of research; Lincoln University who provided a doctoral research scholarship and funding for the project; and our family and friends who supported us as we climbed over mountains.

References

Australian Capital Territory Insurance Authority (ACTIA) (2004) *Guide to Risk Management: Insurance and Risk Management Strategies. Australian/New Zealand Standard AS/NZS 4360:2004* (Canberra: ACT Insurance Authority).

Bachofer, M. (2009) *FCMappers*. Available at http://www.fcmappers.net/joomla/ (accessed 26 March 2010).

Bertsch, V., Treitz, M., Gelderman, J., & Rentz, O. (2007) Sensitivity analyses in multi-attribute decision support for off-site nuclear emergency and recovery management, *International Journal of Energy Sector Management*, 1(4), pp. 342–365.

Buech, F. (2008) Seismic response of Little Red Hill—towards an understanding of topographic effects on ground motion and rock slope failure, PhD thesis, University of Canterbury, 2008.

Conger, J. A., & Kanungo, R. N. (1988) The empowerment process: Integrating theory and practice, *Academy of Management Review*, 13(3), pp. 471–482.

Corbacioglu, S., & Kapucu, N. (2006) Organizational learning and self-adaptation in dynamic disaster environment, *Disasters*, 30(2), pp. 212–233.

Denzin, N. K. (1978) The logic of naturalistic inquiry, in: N. K. Denzin (Ed) *Sociological Methods: A Source Book*, pp. 6–45 (New York: McGraw-Hill).

Dorcey, A. H. J., & McDaniels, T. (1999) Great expectation, mixed results: trends in citizen involvement in Canadian environmental governance. Available at http://www.interchg.ubc.ca/dorcey/trends.

Galley, I., Leonard, G., Johnston, D., Balm, R., & Paton, D. (2004) The Ruapehu lahar emergency response plan development process: An analysis, *The Australasian Journal of Disaster and Trauma Studies*, 2004(1). Available at http://www.massey.ac.nz/~trauma/issues/2004-1/galley.htm (accessed 1 May 2010).

GeoNet (2009) Article: July 16 2009—Fiordland quake biggest for 8- years (updated Aug 5, 1:20 pm). Available at http://www.geonet.org.nz/news/archives/2009/article-jul-16-2009-fiordland-quake-biggest-for-80-years.html (accessed 1 May 2010).

Gibson, C. (2007) HB 167: 2007. *Handbook: Security Risk Management*. (Sydney: Standards Australia; Wellington: Standards New Zealand).

Greene, J. C., Caracelli, V. J., & Graham, W. F. (1989) Toward a conceptual framework for mixed-method evaluation and designs, *Educational Evaluation and Policy Analysis*, 11, pp. 255–274.

International Organization for Standardization (2009) *Risk Management: Principles and Guidelines* ISO/IDIS 31000:2009.

Jick, T. D. (1979) Mixing qualitative and quantitative methods: Triangulation in action, *Administrative Science Quarterly*, 24, pp. 602–611.

Kingsbury, P. (Ed) (1996) *Earthquake Induced Slope Failure Hazard Wellington—Notes to Accompany* Publication WRC/pp-T-95/06, Wellington Regional Council.

Kosko, B. (1987) Adaptive inference in fuzzy knowledge networks, in: *Proceedings of the First IEEE International Conference on Neural Networks (ICNN-8)*, June, San Diego, CA, pp. 261–268.

Kosko, B. (1992) *Neural Networks and Fuzzy Systems: A Dynamical Systems Approach to Machine Intelligence* (Englewood Cliffs, NJ: Prentice-Hall).

Kvale, A. (1953) Seismic seiches in Norway and England during the Assam earthquake of August 15, 1950, *Bulletin of the Seismological Society of America*, 45, pp. 93–113.

Kvale, S. (1996) *Inter Views: An Introduction to Qualitative Research Interviewing* (Thousand Oaks, CA: Sage Publications).

Laszlo, E., Artigiani, R., Combs, A., & Csanyi, V. (1996) *Changing Visions, Human Cognitive Maps: Past, Present, and Future* (Westport, CT: Praeger).

Leonard, G. S., Johnston, D. M., Christianson, A., Becker, J., & Keys, H. (2008) Developing effective warning systems: ongoing research at Ruapehu volcano, New Zealand, *Journal of Volcanology and Geothermal Research*, 172(3–4), pp. 199–215.

McGarr, A., & Vorhis, C. (1968) *Seismic Seiches from the March 1964 Alaska Earthquake. Effects on the Hydrological Regimen*, US Geological Survey. Available at http://www.dggs.alaska.gov/webpubs/usgs/p/text/p0544e.PDF (accessed 6 May 2010).

McIvor, D., & Paton, D. (2007) Preparing for natural hazards: Normative and attitudinal influences, *Disaster Prevention and Management*, 16(1), pp. 79–88.

Mileti, D. (1999) *Disasters by Design: An Re-assessment of Hazards in the United States* (Washington, DC: Joseph Henry Press).

Nakamura, K., Iwai, S., & Sawaragi, T. (1982) Decision support using causation knowledge base, *IEEE Transaction on Systems, Man, and Cybernetics*, SMC-12, pp. 756–777.

New Zealand Ministry of Civil Defence & Emergency Management (2002a) *The Civil Defence and Emergency Management (CDEM) Act 2002*. Available at http://www.civildefence.govt.nz/memwebsite.nsf/wpg_url/for-the-cdem-sector-cdem-act-2002-index?opendocument (accessed 18 April 2010).

New Zealand Ministry of Civil Defence & Emergency Management (2002b) Introduction, *The Civil Defence and Emergency Management Act 2002*. Available at http://www.civildefence.govt.nz/memwebsite.nsf/wpg_url/for-the-cdem-sector-cdem-act-2002-index?opendocument (accessed 18 April 2010).

Ostrom, E. (1999) Coping with tragedies of the commons, *Annual Review of Political Science*, 2, pp. 493–535.

Otway, P. M., Hodgson, K. A., & Narin, I. A. (1995) Whakapapa skifields larhar study, *Science Report*, 95(39), pp. 1–90 (Wellington: Institute of Geological and Nuclear Sciences).

Owens, I., Irwin, D., & McQueen, W. (2002) *Avalanche Accidents in Aotearoa—New Zealand* (Wellington: New Zealand Mountain Safety Council).

Ozesmi, U. (1999) Ecosystems in the mind: Fuzzy cognitive maps of The Kizilirmak Delta Wetlands in Turkey, PhD thesis, University of Minnesota.

Ozesmi, U., & Ozesmi, S. L. (2003) A participatory approach to ecosystem conservation: Fuzzy cognitive maps and stakeholder group analysis in Uluabat Lake, Turkey, *Environmental Management*, 31(4), pp. 518–531.

Ozesmi, U., & Ozesmi, S. L. (2004) Ecological models based on people's knowledge: A multi-step fuzzy cognitive mapping approach, *Ecological Modeling*, 176(2004), pp. 43–64.

Patton, M. (1987) *How to Use Qualitative Methods in Evaluation* (Newbury Park, CA: Sage Publications).

Pearce, L. (2002) Disaster management and community planning, and public participation: How to achieve sustainable hazard mitigation, *Natural Hazards*, 28, pp. 211–228.

Pelling, M. (2007) Learning from others: The scope and challenges for participatory disaster risk assessment, *Disasters*, 31(4), pp. 373–385.

Reed, M. S. (2008) Stakeholder participation for environmental management: A literature review, *Biological Conservation*, 141(10), pp. 2417–2431.

Reimann, S. (1998) On the design of artificial auto-associative neural networks, *Neural Networks*, 11, pp. 611–621.

Saunders, W., & Glassey, P. (Eds) (2007) *Guidelines for Assessing Planning, Policy and Consent Requirements for Landslide-prove Land* GNS Science Miscellaneous Series 7, Wellington, New Zealand.

Saunders, W., Forsyth, J., Johnston, D., & Becker, J. (2007) Acting accordingly: Using the RMA & CDEM Act to improve community resilience. Paper presented at the New Zealand Planning Institute 'Politics and Planning' Conference, 28–30 March. Available at http://www.gns.cri.nz/services/hazardsplanning/downloads/saundersforsythwithfigures.pdf (accessed 15 March 2010).

Serway, R. A., & Jewett, J. W. (2006) *Principles of Physics: A Calculus Based Text*, 4th ed. (Toronto: Thompson).

Sharpe, S. C. F. (1960) *Landslide and Related Phenomena: A Study of Mass-Movements of Solid and Rock* (New Jersey: Pageant Books).

Spreitzer, G. (2007) Taking stock: A review of more than twenty years of research on empowerment at work, in: C. Cooper & J. Barling (Eds) *The Handbook of Organizational Behaviour*, pp. 1–42 (New York: Sage Publications).

Strickert, G. E., Samarasinghe, S., & Davies, T. (2009) Resilience models for New Zealand's Alpine Skiers based on people's knowledge and experience: A mixed method and multi-step fuzzy cognitive mapping approach. Presented at MODSIM 09, Cairns, Australia, 18 July.

Tashakkori, A., & Teddlie, C. (1998) *Mixed Methodology: Combining Qualitative and Quantitative Approaches* Applied Social Research Methods Series (New York: Sage Publications).

The Resilience Alliance (2007) *Adaptive Capacity*. Available at http://www.resalliance.org/565.php (access 22 July 2007).

Thomas, K. W., & Velthouse, B. A. (1990) Cognitive elements of empowerment: An 'interpretive' model of intrinsic task motivation, *Academy of Management Review*, 15(4), pp. 666–681.

Trucco, P., Cagno, E., Ruggeri, F., & Grande, O. (2008) A Bayesian belief network modeling of organizational factors in risk analysis: A case study in maritime transportation, *Reliability Engineering & Safety Systems*, 93(6), pp. 845–856.

Varnes, D. J. (1978) Slopem types and processes, in: R. L. Schuster & R. J. Krizek (Eds) *Special Report 176: Landslides Analysis and Control*, pp. 12–33 (Washington, DC: TRB, National Research Council).

Walker, B., Carpenter, S., Anderies, J., Abel, N., Cumming, G. S., Janssen, M., Lebel, L., Norberg, J., Peterson, G. D., & Pritchard, R. (2002) Resilience management in social-ecological systems: A working hypothesis for a participatory approach, *Conservation Ecology*, 6(1), p. 14. Available at http://www.ecologyandsociety.org/vol6/iss1/art14 (accessed 10 July 2006).

Wegner, E., McDermott, R. A., & Snyder, W. (2002) *Cultivating Communities of Practice: A Guide to Managing Knowledge* (Boston, MA: Harvard Business School Publishing).

Weichselgartner, J. (2001) Disaster mitigation: The concept of vulnerability revisited, *Disaster Prevention and Management*, 10(2), pp. 85–94.

Yetton, D., Wells, A., & Traylen, N. J. (1998) *Probability and Consequences of the Next Alpine Fault Earthquake* (Christchurch, New Zealand: Geotech Consulting).

Appendix A. Chart Displaying Simulation Results of Policy Considerations for TSGFCM.

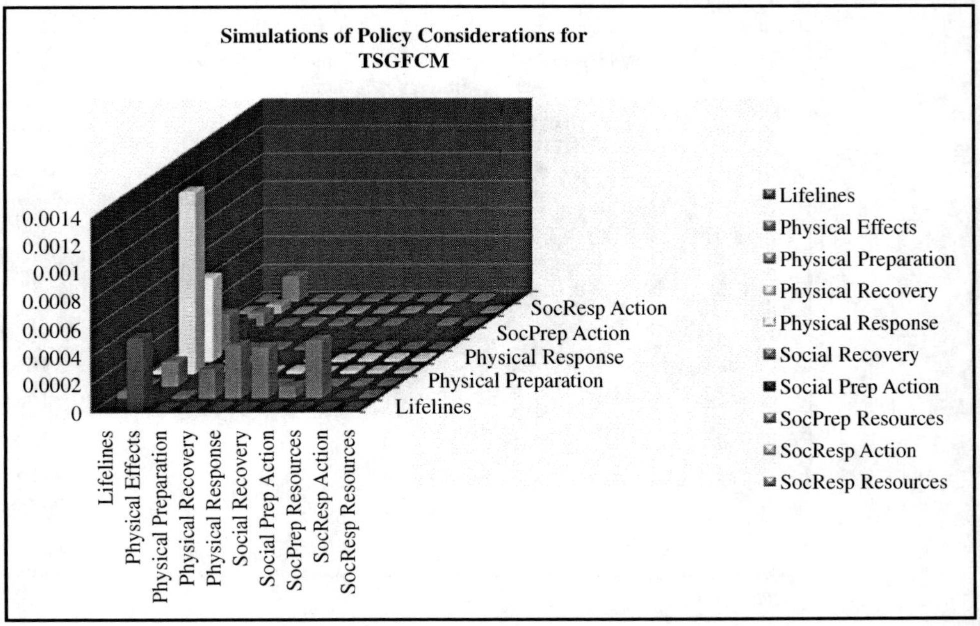

Appendix B. Chart Displaying Simulation Results for Policy Considerations for SAM.

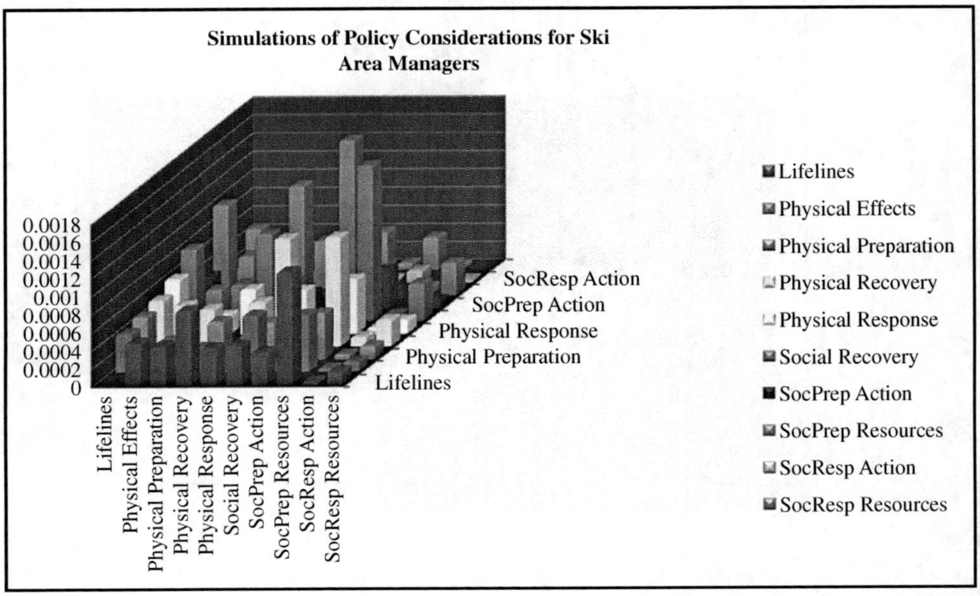

Appendix C. Chart Displaying Simulation Results for Policy Considerations for SIW.

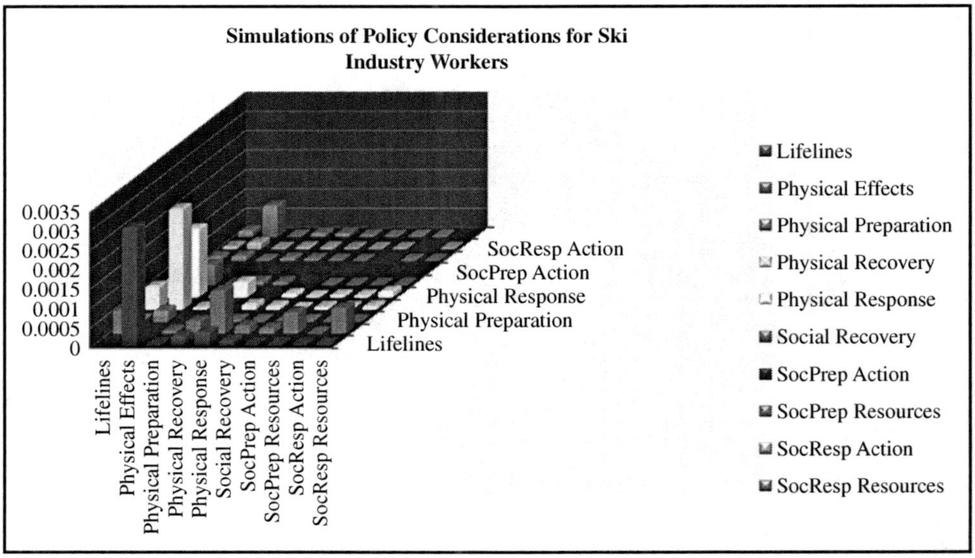

Appendix D. Chart Displaying Simulation Results for Policy Considerations for ASR.

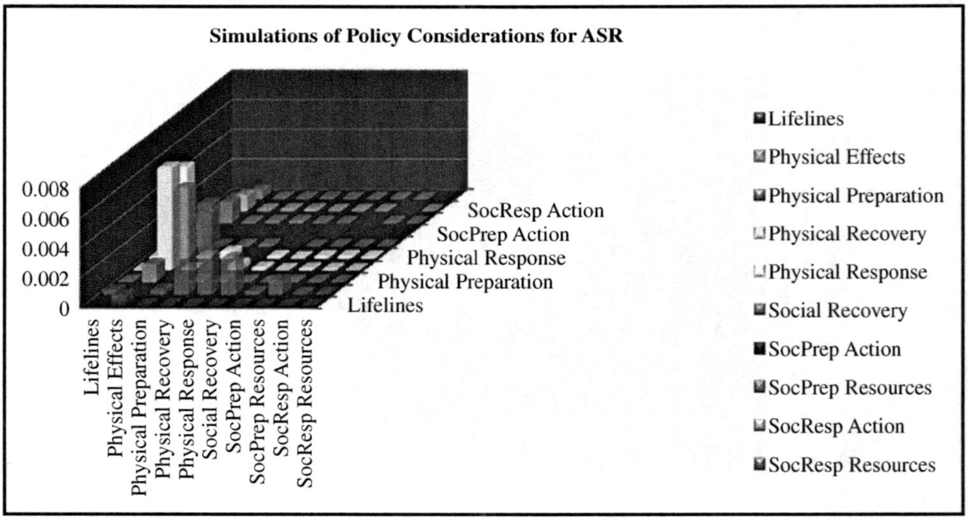

Index

Note: Page numbers in **bold** type refer to figures
Page numbers in *italic* type refer to tables
Page numbers followed by 'n' refer to notes

Abruzzo (Italy) *see* L'Aquila earthquake (2009) 9
aggregate shocks 40–9
agriculture 31–3, 36, 46; farming 26, 29–35
aid 1, 3, 21–2, 42–3, 46–9
Albala-Bertrand, J.M. 43
Alexander, D.E. 2–3, *5*, 8–25
Allen, T. (National Incident Commander) 65
Alpine Snow Riders (ASR) 79, 83–5, *83*, *90*
Asia/Indian Ocean earthquake/tsunami (2004) 32, 46
Assam (Medog) earthquake (1950) 80
assets 4, 40, 45–6; smoothing 3, 41
Augustus System 12
Australia 75; Capital Territory Insurance Authority (ACTIA) 75
Auto-Associative Neural Networks (AANN) 75–77
avalanches 74–6, 80–1
Avezzano earthquake (1915) 10, 20

Balgah, R.A.: and Buchenrieder, G. 3–4, *5*, 40–53
Ban Ki-Moon (UN Secretary-General) 1
Banzhaf, S.: and Boyd, J. 37n
Barberi, F. 13
Bayesian belief networks 76
Bazerman, M.H. 35
belief networks: Bayesian 76
Berlusconi, S. (Italian Prime Minister) 15, 19
Block, R.: and Cooper, C. 23
Boston: flooding (1968) 33–4
Boxer, B. (Senator) 34
Boyd, J.: and Banzhaf, S. 37n
British Petroleum (BP) 65, 68; Gulf Coast/ Deepwater Horizon oil spill (2010) 4–6, 54–5, 63–4, 67
Browning, Iben: earthquake prediction 22
Buchenrieder, G.: and Balgah, R.A. 3–4, *5*, 40–53
Burma: Nargis cyclone (2008) 1

Calvi, G.: and Spaziante, V. 16
Catastrophic Natural Events (CNE) 84–5
Catskill-Delaware watershed 34
Charles River: watershed 26, 30, 37n
Chemical, Biological, Radiological and Nuclear (CBRN) defence 11
China: Sichuan earthquake (2008) 1
Cialente, M. (Mayor of L'Aquila) 16, 20
civil defence/protection 3, 8, 11–16, 21–2
climate change 2, 6, 26
Collins, F. 70
Columbia: Nevado del Ruiz eruption (1985) 22
Common Reference Frame (CRF) **56**, 57
communication 12, 84
community 2–4, 26, 66; coalition 36; mobilization 47; protection 33; support 4, 40
conceptual framework 72, 86
conservation 17, 26–33; land 17, 26–8
consumption: smoothing 3, 41; variance 41
contamination 26–37, *see also* pollution
Cooper, C.: and Block, R. 23
Corexit 67, 68, 69
Cossiga, F. (Italian President) 11
covariate shocks 4, 40–50
Cronton River: watershed 34
cyclones 1; Nargis (Myanmar/Burma 2008) 1

dams 28–30
decision-making 54–8, **56**, 63, 70; democratic 3, 8, 23
Deepwater Horizon oil spill (2010) 4–6, 54–5, 63–4, 67
deforestation 2, 29
democracy 3, 8, 23
developing countries 29, 41, 44, 50
Dickson, D. 36
disaster 22, 33–4, 37, 54–5, 72–3, 78–80; awareness 43, 50, 72, 84–6; impacts 2, 32, 84
disease 2, 26, 29, 42, 47, 66
Dorcey, A.: and McDaniels, T. 73
Dovers, S.: and Handmer, J. 8

Drama Theory II (DT II) 4, 54–70, **56**; build-up 56–7, **56**; dilemmas 54–9, **58**, **60**, **61**, **62**, 64, 70; disaster management case study 63–6; doubts 54–6, 59–63, **60**, **61**, **62**; foundations for disaster management 55–7; intentions 54–63, **60**, **61**, **62**; positions 54–63; scene-setting 54–7, **56**
Draquila: Italy Shakes documentary 10
drinking water 26–37

earthquakes 1–5, 8–23, **18**, 41, 72–86, **79**; Asia/Indian Ocean (2004) 32; Assam (1950) 80; Avezzano (1915) 10, 20; Fiordland (2009) 74; Haiti (2010) 1, 10; Iben Browning prediction (1990) 22; Irpinia (1980) 11, 21; L'Aquila (2009) 3, 8–23, **10**, **15**, **18**; Mexico City (1985) 23; Molise (San Giuliano di Puglia)(2002) 12, 22; Pakistan (2005) 10; post-earthquake housing 16–21, **17**, **18**, **19**, **20**; Sichuan (2008) 1; Strait of Messina (1908) 10, 20–1
economic development 27–9, 33–4, 42
economic impacts 1, 6, 11, 19–20, 27–8, 40, 43, 55
ecosystem 6, 26–32, 45, 50, 67–9, 76; services 28–29, 32, 37n
Eisinger, P. 23
Emergency Operations Centres (EOCs) 12, 15; Command and Control (DICOMAC) 12, 15; Local Operations Centre (COC) 12; Mixed Operations Centre (COM) 12, 15; Permanent Unified Operations Room (SOUP) 12
employment 11, 17, 20, 29
environmental damage 26, 54–5, 65–7
Estes-Smargiassi, S. 29
evacuation 16, **17**, 21, 49
Exxon Valdez oil spill (1989) 68

Farber, S. 30
farming 26, 29–35
Fiordland earthquake (2009) 74
floods 26–30, 41, 48–9; Boston (1968) 33–4; control 26–7, 33–7, 37n
Friends of the Napa River (community coalition) 33–6
Fuzzy Cognitive Maps (FCM) 75–86, **79**, *83*

game theory 54–7, **56**, 70
gap hazard analysis 4, 72–86
geomorphic assessment 5, 76–85
Giolitti, G. (Italian Prime Minister) 21
Giuliani, G.G. 12–13
Gopalakrishnan, C.: and Levy, J.K. 1–7
government 26–35, 37n, 49, 65, 68; policy 3, 8–23
Greenville watershed 28
Guide to Risk Management (Australia) 75
Gulf Coast 1, 4, 54–70; /Deepwater Horizon oil spill (2010) 4–6, 54–5, 63–4, 67; hurricane (2005) 68; tribes 54–70
Günther, I.: and Harttgen, K. 41
Guzzanti, S. 9–10

Haiti earthquake (2010) 1, 10
Handmer, J.: and Dovers, S. 8
Harrah's Entertainment 30
Harttgen, K.: and Günther, I. 41
health 2, 6, 11, 27, 35, 47, 66–9, 73
homeless 2–3, 8–9, 17, 20
Horizon Oil Spill Science and Engineering Review Team 65
housing: post-earthquake 16–21, **17**, **18**, **19**, **20**
hurricanes 1, 22–3, 29, 32–4, 34, 46–9, 49, 65–70; Andrew (1992) 32; Gulf Coast (2005) 68; Jeanne (2004) 29; Katrina (2005 1, 22–23, 46, 49, 68–69; Mitch (1998) 46–48; Stan (2005) 48
Hussein, S. 14

Il Centro 13
India: Assam (Medog) earthquake (1950) 80
Indian Ocean earthquake/tsunami (2004) 32, 46
Informationally Closed Environment (ICE) 56
Integrated Pest Management (IPM) 33
Irpinia earthquake (1980) 11, 21
Italy 1, 3, 8–23, **10**, **15**, **18**; Avezzano earthquake (1915) 10, 20; Chemical, Biological, Radiological and Nuclear (CBRN) defence 11; Commission on Major Risks (*Commissione sui Grandi Rischi*) 13–14; Decree-Law (DL) 21; Democratic Party 16; Department of Civil Protection (DCP) 12–13, 16, 21; financial police force (*Guardia di Finanza*) 15; government policy on disaster response 8–23; Intelligence and Democratic Security Service (SISDE) 11; Irpinia earthquake (1980) 11, 21; L'Aquila earthquake (2009) 3, 8–23, **10**, **15**, **18**; Ministry of Civil Protection 11; Ministry of Interior 11; Molise earthquake (2002) 12, 22; National Fire and Rescue Service 11; National Institute for Astrophysics (INAF) 13; Organization for Economic Co-operation and Development (OECD) 12; People of Freedom 19; Red Cross 15; San Salvatore hospital 9; State for Civil Protection 13; Strait of Messina earthquake (1908) 10, 20–1; Sustainable and Eco-friendly Anti-seismic Complex (CASE) 16–20, **18**, **19**

Jackson, L. 67–9
Jacoby, H.: and Skoufias, E. 48
Jewett, J.W.: and Serway, R.A. 80
Jindal, B. (Governor) 63–5, **64**
Johnston-Block, M. 33–6

Kingdon, J. 33
Kirkham, C. 65
Kirschenbaum, A. 21
Kousky, C. 3, *5*, 26–39
Kvale, A. 80

Laffon, B. 68
land 2, 30; acquisition/ownership 27, 30–2, 36, 46; city-owned 28; conservation 17, 26–8
land-use 27, 30–1, 37n
landslides 26, 29, 32, 76–80
L'Aquila earthquake (2009) 3, 8–23, **10**, **15**; post-earthquake housing 16–21, **17**, **18**, **19**, **20**
Lavell, A. 23
Lee, A. (Colonel) 65
Leopold, L. 33
levees 28–9, 36, 66
Levine, A. 69–70
Levy, J.K. 4, *5*, 54–71; and Gopalakrishnan, C. 1–7
Louisiana 54–70; Department of Health and Hospitals 69; Spirit 69, *see also* Gulf Coast

McDaniels, T.: and Dorcey, A. 73
man-made disaster 3, 6, 40–1, 50
management: disaster *5*, 8–16, 26, 35–6, 40–50, 54–76, 84–6; risk 3–4, 40–50, 73, 75
mangroves 26, 29, 32–5
Marine Management Organization (MMO) 67
marshlands 55, 66
Masey, D. 74
Messina Strait: earthquake/tsunami (1908) 10, 20–1
Mexico City earthquake (1985) 23
microeconomic theory 40–3, 49–50
Mileti, D. 73
Millennium Ecosystem Assessment (2005) 37n
Mitchell, T.: *et al* 48–9
mitigation *5*, 8–16, 26, 35–6, 40–50, 54–76, 84–6
Molise (San Giuliano di Puglia) earthquake (2002) 12, 22
Myanmar: Nargis cyclone (2008) 1

Nalco Holding Company 68
Nagaris cyclone (Myanmar/Burma 2008) 1
natural disaster 1–3, 6–8, 40–2, 47, 50, 73–4, 78, 81, 86
Nestlé Waters 31
Nevado del Ruiz volcanic eruption (1985) 22
New Zealand 4–5, 72–5, 79; Alpine Snow Riders (ASR) 79, 83–5, *83*, *90*; Civil Defence Act (1983) 74; Civil Defence and Emergency Management (CDEM) Act (2002) 74–5, 86; Earthquake and Geological Hazard Mitigation Strategy 75; Fiordland earthquake (2009) 74; Human Ethics Committee 79; Regional Policy Statements 75, 86; Resource Management Act (RMA)(1991) 74, 86; Ski Area Managers (SAM) 79, 84, *89*; ski areas 72–5; Ski Industry Workers (SIW) 79, 82–4, *83*, *90*; Total Social Group FCM (TSGFCM) 83–5, *89*
Non-Governmental Organizations (NGOs) 6, 11, 47–8

Obama, B. (US President) 65
oil spills 4–6, 54, 54–5, 63–9, 65–9; Exxon Valdez (1989) 68; Gulf Coast/Deepwater Horizon (2010) 4–6, 54–5, 63–4, 67; Prestige (2002) 68
Organization for Economic Co-operation and Development (OECD) 12, 44; report (2010) 12, 23
Ott, R. 68
Oxfam 47
Ozesmi, U. 76; and Ozesmi, S. 75, 78

Pakistan: earthquake (2005) 10
Parker, S.: and Skoufias, E. 48
Participatory Disaster Risk Assessments (PDRA) 73
Payments for Ecosystem Services (PES) 31–2
Pearce, L. 73
Pelling, M. 73
Plan International 47–49
policy 8–10, 26, 33, 37, 55; adoption 27; change 33; development with mountainous communities 72–86; government 3, 8–23; initiatives 72; setting 29, 32, 78, 85
politics 2–3, 9–10, 21, 26–33, 37
pollution 12, 26–9, 35–7; water 26, 29–33, 36
post-earthquake housing 16–21, **18**, **19**, **20**; shelter **17**
poverty 3, 42–4, 49
prediction 8, 14, 22, 72
Prestige oil spill (2002) 68
prevention 2, 26–38
public goods 31, 35, 41
public intervention 43–4
public participation 73

qualitative interviews 75–85
Quasi-Autonomous Non-Governmental Organization (QUANGO) 11

receiver variables 82, *83*
reconstruction 16, **17**, 19, 43, 46–7, 84
recreation 26–30, 34
Red Cross: Italy 15; Thailand 47
redevelopment **17**
Redlener, I. 68
Reed, D. 65
rehabilitation 43
rehousing 20
relief 1, 21–2, 42–3, 46–9
relocation 2, 18, 66
rescue 11–12
reservoirs 80, **80**
resettlement **17**, 21, 45, *see also* post-earthquake housing
resilience/recovery 5–8, *5*, 14–23, 40–50, 70–4, **79**, 85–6; informal instrumental 46–9; local 16
risk: analysis 2, 30, 73; reduction 28–32, 37n, 45
risk management 45–50, 73, 75; instruments 3–4, 40–2, 45, 48–9; social 42–6
Robichaux, B.D. 66
rock fall 76, 80–1
Ruapehu volcanic eruptions (1969 and 1975) 74
Ruefer, J. 30

San Salvatore hospital (Italy) 9
search and rescue 11–12
seismic activity 9, 12–14, 22, 80, *see also* earthquakes
Serway, R.A.: and Jewitt, J.W. 80
severe weather 78, 81, *see also* storms
Shaw Group 65
shelter: post-disaster 16, **17**
Sichuan earthquake (2008) 1
Ski Area Managers (SAM) 79, 84, *89*
ski areas: New Zealand 72–5
Ski Industry Workers (SIW) 79, 82–4, *83*, *90*
Skoufias, E. 50; and Jacoby, H. 48; and Parker, S. 48
Social Group Fuzzy Cognitive Maps (SGFCM) 77, 82, 86
Social Preparation Action (SPA) *82*
Social Preparation Resources (SPR) *82*, 84–5
social recovery *82*, *83*, 84–5, *89*, *90*
Social Response Action (SRA) *82*
Social Response Resources (SRR) *82*, 85
Spaziante, V.: and Calvi, G. 16
stakeholders 29, 32–7, 64, 72–3, 77, 86
state intervention 20, 40
storms 26, 29, 32–4; damage 26, 29, 32–4
Strait of Messina earthquake/tsunami (1908) 10, 20–1
Strickert, G.: *et al* 4, *5*, 72–90, *82*, *83*
Sturzstroms 76, 80
Surface Water Treatment Rule (1989) 34
sustainability 5, 44, 50–4, 72–3, 86
Sustainable and Eco-friendly Anti-seismic Complex (CASE) 16–17, 20; post-earthquake housing **18**, **19**
Sustainable Hazard Mitigation (SHM) 5, 72–5, 86
Suttles, D. 69

technological disaster 2, 6
Thailand 43; Lampoon community 43; Red Cross Society 47
Total Social Group FCM 83–5, *89*
tourism 17, 29–30, 46–7, 78
transmitter variables 82–3, *83*
transport 12, 84
triangulated approach 75–8
Trosper, R.L. 48
Truckee River (Reno, Nevada) 27–9
tsunamis 10, 32, 46; Asia/Indian Ocean (2004) 32, 46; Strait of Messina (1908) 10

United States of America (USA) 3–4, 23, 26; Army Corps of Engineers 28–30, 34–7, 37n, 63–5, **64**; Centers for Disease Control (CDC) 68; Department of Homeland Security 65; Department of the Interior 37n; Emergency Support Functions (ESFs) 12; Environmental Protection Agency (EPA) 28, 35, 65, 67–9; Federal Emergency Management Agency (FEMA) 12, 49, 70; Federal Government 63–6, **64**; Greenville Water System (GWS) 28; Gulf Coast tribes 54–70; Gulf Coast/Deepwater Horizon oil spill (2010) 4–6, 54–5, 63–4, 67; Homeland Security Presidential Directive-5 (HSPD-5) 65; House Subcommittee on Insular Affairs Wildlife and Oceans 66; Memorandum of Agreement (1997) 35; Metropolitan District Commission 31; National Center for Disaster Preparedness 68; National Commission on Children and Disasters 68; National Institutes of Health (NIH) 70; National Oceanic and Atmospheric Administration (NOAA) 65; National Response Framework (NRF) 65; Nature Conservancy 28, 37n; Safe Chemicals Act (2010) 69; Substance Abuse and Mental Health Services Administration (SAMHSA) 70; Toxic Chemicals Safety Act (2010) 69; Toxic Substances Control Act (TSCA)(1976) 69; Watershed

Protection Act (1922) 31, *see also* Gulf Coast; Louisiana

Vittel 31–2, 35
volcanic eruptions 41, 74; Nevado del Ruiz (1985) 22; Ruapehu (1969 and 1975) 74
volunteer organizations 11, 15–16, 22, 67

water 26–37, 48; contamination/pollution 26, 29–33, 36; drinking 26–37; filtration 26–9, 34
watersheds 26–36, 37n; Catskill-Delaware 34; Charles River 26, 30, 37n; Cronton River 34; Greenville 28
weather: severe 78, 81, *see also* storms
Weichselgartner, J. 73
welfare 41–45, 50
wetlands 26–36, 66
World Bank 44; Social Risk Management framework 44–6, 50
World Vision 47
Wunder, S. 31